- Starting On Pole -

The New Professional's Guide to the Motorsport Industry

Tom Roberts

Copyright © 2016 Torotex Engineering

All rights reserved.

ISBN: 978-1539418238

Cover design by Luko@99designs

"Racing is life. Everything before or after is just waiting."

– Steve McQueen

DISCLAIMER

This document and all contents therein are the sole property of Torotex Engineering. This book is intended for educational and references purposes only, and is not to be considered definitive. Opinions expressed are that of the author unless otherwise stated and do not necessarily reflect the views and opinions of Torotex Engineering. Any likeness or similarity to other publications is purely coincidental, and all precautions possible have been taken to ensure firm plagiarism and copyright parameters have been adhered to.

This book may not be replicated in part or in full without the express and written consent of Torotex Engineering. Replication, disclosure or sharing of this publication to any third parties shall be deemed a copyright infringement.

Acknowledgements

This book could not exist without the help, guidance and advice of many people. Both close friends and professional colleagues have contributed to its coming to fruition and I cannot possibly list every influence I have had in writing it.

My family and friends for their continued support throughout this project, enduring the stress, anxiety and lack of a social life alongside me in order to get this work to the printers; thank you all.

To the colleagues I have met and had the pleasure and privilege of working with in the last few years; many of you are now great friends. The opportunities granted to me which allowed this book to exist would not have been possible without you. Gummi, your input based on years of your own experience was invaluable in piecing together the traits of a successful engineer.

Steve, your continued support throughout this project has been fantastic and without the push I received from you, I would not be where I am today.

Tim, you initiated the drive to get this going. You're a great inspiration.

Thank you all.

TABLE OF CONTENTS

INTRODUCTION .... - 1 -
ABOUT ME .... - 1 -
ABOUT THIS BOOK .... - 3 -

THE GOOD, THE BAD, AND THE SURPRISING .... - 7 -
THE WORST JOB IN THE WORLD? .... - 8 -
THE BEST JOB IN THE WORLD? .... - 10 -

GETTING STARTED .... - 15 -
THE FUNDAMENTALS .... - 16 -
FORMAL EDUCATION .... - 18 -
Degrees .... - 18 -
Apprenticeships .... - 26 -
SPECIALIZING .... - 27 -
Aerodynamics .... - 28 -
Powertrain .... - 30 -
Chassis, Brakes and Suspension .... - 32 -
Electronics .... - 33 -
Hybrid Technology .... - 35 -
Data Engineering .... - 36 -
Fabricators .... - 38 -
Composite Engineers .... - 39 -
Grid Support Role .... - 41 -
PROFESSIONAL BODIES .... - 45 -
PERSONAL FITNESS .... - 47 -

LANDING YOUR FIRST JOB .... - 51 -

THAT ILLUSIVE EXPERIENCE	- 52 -
SOCIAL MEDIA	- 55 -
YOUR RÉSUMÉ	- 56 -
ONLINE TESTING	- 60 -
THE INTERVIEW	- 61 -
Academics	*- 63 -*
Experience	*- 64 -*
Personal	*- 65 -*
Phone Interviews	*- 66 -*
ASSESSMENT CENTERS	- 67 -

THE NITTY GRITTY — - 71 -

ATTITUDE	- 72 -
EFFICIENCY	- 74 -
COMMUNICATION	- 76 -
ORGANIZATION	- 80 -
KNOWING YOUR ROLE	- 84 -
CONFIDENTIALITY	- 87 -
THE COMPETITION ARE YOUR COLLEAGUES	- 89 -
HEALTH AND SAFETY	- 90 -

TEAMWORK — - 95 -

THICK SKIN	- 96 -
MANAGERS	- 98 -
PEERS	- 99 -
SUBORDINATES	- 101 -
DRIVERS	- 102 -
Confidence in the Car	*- 103 -*
The Red Mist	*- 104 -*

Amateur Drivers .... - 105 -
Professional Drivers .... - 107 -
Gentleman Drivers .... - 110 -
PROFESSIONAL RELATIONSHIPS .... - 111 -

OTHER THINGS TO KNOW .... - 113 -

FOOD AND DRINK .... - 114 -
CLOTHING .... - 115 -
TRAVEL .... - 116 -
ADDITIONAL EQUIPMENT .... - 119 -
REPRESENTING YOUR TEAM .... - 118 -
DEALING WITH THE MEDIA .... - 119 -
PARC FERMÉ .... - 124 -
RAW CARROTS .... - 124 -

NETWORKING .... - 125 -

WHO YOU KNOW .... - 126 -
WHO KNOWS YOU .... - 127 -
WHAT YOU KNOW .... - 129 -
WHAT THEY KNOW YOU KNOW .... - 130 -
YOUR REPUTATION .... - 132 -

GOING SOLO .... - 135 -

CASH FLOW .... - 136 -
IMAGE .... - 137 -
KEEPING ON TOP OF THINGS .... - 139 -
MARKETING .... - 141 -

SO, IN CONCLUSION .... - 143 -

RECOMMENDED READING .... - 147 -

Chapter One

INTRODUCTION

"I am not bound to win, but I am bound to be true."
– Abraham Lincoln

About Me

When I started out working in motorsport, I got a huge shock. No amount of education, coursework, research or guesswork could have prepared me for the true nature of the industry.

My first motorsport job was volunteer work for a small race team based in Essex in the UK. I was 16 when I first got to be around one of their racing cars and it felt like it was going to be the start of a high flying and fast paced career in an industry I grew to love. This was by no means my first experience of working with cars, but it was my introduction to doing so competitively.

I achieved reasonably good grades throughout school, although I am the first to admit I took my foot off the gas (excuse the pun) during my college years. I was fortunate enough to be accepted into a respected and high achieving university, offering a degree in something that I was really interested in; motorsport engineering.

My luck continued straight out of university where I landed a job as an Application Engineer for Cosworth. This role opened a wealth of opportunities and experience to me, allowing me travel all over the world and compete with teams in countless disciplines of motorsport. I worked on ground-up projects in multiple different touring car championships, endurance racing, motorcycle racing, hatchback series, world rally cross, formula racing and even powerboats to name a few. This fortuitous opportunity gave a breath-taking insight into the varied world of motorsport.

Upon leaving Cosworth, I began working as a contractor under my own company, something I recommend everyone tries during their career. This allowed me to continue doing the job I loved whilst furthering myself, my career and my reputation.

About This Book

This book is not intended for me to paint a picture of myself, or the motorsport industry, as perfect. It is intended to show the opportunities that can present themselves to anyone with a lot of hard work… and a little luck.

I write this book from a personal standpoint and draw upon my own experiences. There are anecdotes and inputs from various people I have met along the way, but I do not claim this to be a step-by-step guide on succeeding. Success in motorsport is highly dependent on you as an individual, and will vary for each person who sets out to conquer the industry.

This book simply lays out what I wish I had known upon starting out. The pitfalls and mistakes I have made myself, or seen others make whilst working in motorsport. It lays out ways in which you can help yourself get that first step into an industry that is notoriously impermeable. The book's sections are intentionally succinct and can be considered reference material as and when required.

I come at motorsport from an engineering point of view. I have trained as an automotive engineer, specializing in motorsport, and so this is unfortunately the only fountain of knowledge from which I can draw upon. I do delve into the management techniques I undertook when I began working for myself, but there are far more specific and useful books on running your own company that are detailed at the end.

I want to make clear that engineers are not the only employees of motorsport teams – everyone from caterers, public relations officers, human resources and accountants are all required within any successful team. I would claim however, that to be at the sharp end, engineering is where you want to be. It is the engineers who are on the pit wall during a race as well as deciding on the tactics and strategies to employ in order to secure a victory. Engineers design the cars and make sure those cars finish the race. The engineering side of motorsport will give you the greatest exposure, so that is where I would urge you to look.

I also want to state that motorsport is not just about cars. Car racing is the largest sector within the industry, but it by no means the only one. Motorcycle racing is incredibly popular, especially in Europe and Australia. Boat racing holds its own in some of the more affluent areas of the world. And even aircraft racing has its moment in the spotlight now and then. I apologize in advance for the large emphasis placed on cars, but that is where the majority of my experience has come from. The ideas laid out in this book are transferable between all disciplines within motorsport, so please don't take the word "car" literally when it is written.

I invite you to visit the accompanying website (www.startingonpole.com) which is updated regularly with insights and anecdotes, as well as offering helpful advice and assistance for education and applications. The website is constantly evolving, and there is a newsletter that will keep you up to date.

Finally, I want to wish you good luck. By reading this book it shows you are about to embark on a career that will undoubtedly be exciting, challenging and rewarding. Motorsport is a fantastic way of life, and I encourage everyone to experience it in whatever way they can. I hope that this book goes some way to helping you achieve your dreams and goals, and I wish you the best in your career ahead.

– Tom Roberts

Chapter Two

THE GOOD, THE BAD, AND THE SURPRISING

"There is a point when facing the unknown stops being a longed-for adventure and becomes a terrifying reality."
– Storm Constantine

I draw heavily on my own experience for this chapter to try and dispel some of the myths surrounding the motorsport industry. It is not always what you expect, and it is not always going to put a smile on your face. It will, however, continue to inspire you if you let it.

The Worst Job in the World?

It might seem strange for me to start this book by trying to convince you that motorsport is the worst job in the world. I would agree; that is somewhat counter-intuitive. However, there is method in my madness.

The motorsport industry is tough, unforgiving and relentless. You have to *want* to be in it in order to *survive* being in it. In the process of reading this chapter, I invite you to change your mind. To put this book down and have a rethink about your life's trajectory. Motorsport isn't for everybody, otherwise, let's face it, everybody would be doing it! To the select few who get to the end of this chapter and still have a true desire to tackle this industry, welcome to an exclusive club. The points I make in the next few paragraphs will be revisited during this book, but I want to lay them out for you now.

Motorsport is full of passionate people. I've not met anyone who is in motorsport for any reason other than a passion for it. No one joins for the glamour or money (you'll find out later why) – it is about loving what you do. This passion will definitely result in high tensions and raised tempers if you are pursuing a career in motorsport. You will, at some point, get yelled at. Absolutely no doubt about that.

Motorsport is a 24/7/365 calling. If you cannot commit to that, I would look elsewhere. I have been in conference calls with customers on Christmas day, and spent my birthday alone in a lounge at Bratislava airport with nothing more than boiled sweets to eat. Do not enter the motorsport industry expecting to work your weekends at a race track and take the rest of the week off. Be prepared to give up all of your free time for weeks or months on end. It can take its toll.

Sometimes a race will not go your way. You will find yourself out of contention through no fault of your own, or your driver's. The reasons for not finishing a race are numerous, and when that happens, all the hard work you have put into getting there is wasted. Most crashes happen in the opening lap of a race when the cars are tightly bunched together, so you might not make it to the second corner before your race is over. This is a hard thing to swallow, and you must be able to pick yourself up from these low points and turn things around. Get the car back out for the next race and try again.

The trackside environment can be an extremely stressful place. There is always a huge amount of time and money invested in getting a car (or bike) on to the starting grid. And of course there is the famous quip of "To finish first, first you have to finish". No one wants to see their car fail to finish a race. And even more so, no one wants to be the *reason* for their car not to finish a race. This stress can add up, especially if your team is having a particularly unlucky event or season. You will need to be able to cope with the pressure in a calm and collected manner. Keeping your cool is not always easy, but those who do are the ones who make it.

The Best Job in the World?

So if motorsport is such a tough work environment, why do so many people do it? What's more, why are there so many *other* people that *want* to do it?!

That's an easy question to answer, and I touched on it in the previous section. Passion. People simply love going racing. There is something so exciting and thrilling about the eclectic mix of danger, competition, sounds, sights and smells that anyone in motorsport will be familiar with. Above all else, *this* must be your drive to work in the industry. You must have a fundamental love and passion for racing.

Motorsport has opened a huge number of opportunities for me. I have travelled to places I would never have had thought to. I have worked with some of the most amazing people you will ever encounter. I have looked after some fantastic machinery. I have been at the forefront of the latest technology and innovation. And, I have had the thrill of being an essential cog in securing victories. It is truly a fascinating and exciting industry to be a part of, and something that you will no doubt find out very quickly in your own career.

Notice at this point that there has been no mention of glitz, glamour, partying or money. The reason for that is you cannot join the motorsport industry simply for those reasons. You won't encounter many people along the way who are there for the money. Motorsport is extremely hard work, and there are far easier ways of filling your bank account if that is indeed your end goal.

The parties can be great, and provide an excellent opportunity for networking (more on this later). But the big parties are few and far between. More often than not, before a race you will be working late to get the car ready, and after a race you will be packing up the trucks, debriefing the team, and doing your best to get home for some rest. A quiet drink with members of the team or a team meal in the evening is not unheard of, but take those opportunities to get to know the people you are working with.

Within a few weeks of leaving university, I was sent to look after a twin jet turbine boat in Mississippi. For me, this was just about the best start I could have hoped for. Not because I have any particular love for marine motorsports (it certainly has its perks, though!), more because it was something so different. It was totally unexpected and nothing like what I imagined "motorsport" to be like. It was an opportunity straight off the bat for me to prove myself in a notoriously cut throat industry. It immediately opened my eyes to just how varied the motorsport world was and that lesson has proved invaluable. I urge you to take any opportunity that presents itself – you will always learn something, and will probably thoroughly enjoy yourself in the process!

I finish this chapter with a quote from a lecturer I had whilst at university. He was one of the few lecturers who had actually pursued a career at the frontline of motorsport and supported cars and teams as an engineer during events.

"If you are part of a team that wins, you can rightfully claim that 'today, I am the best in the world at my job. No one did my job better than me today'."

– Jeff Peters

That quote stuck with me throughout university, and then once I experienced a win, I knew exactly what it meant. There is nothing quite like that feeling, and it reaffirmed for me that motorsport was the life I wanted to pursue.

Chapter Three

GETTING STARTED

"The act of taking the first step is what separates the winners from the losers."

– Brian Tracy

In days gone by, motorsport had its fair share of charlatans. A basic level of understanding of one area of a car often opened a door to the entire industry, and a "fake it til you make it" attitude carried you to a certain extent. If they didn't learn on the job FAST, it was often that they were discovered and soon shown the way out.

Nowadays, you simply cannot cheat your way in.

The Fundamentals

I was 4 years old when I first remember helping my father fix a 1976 Triumph Dolomite. I say "helping", I expect I was a nuisance and a hindrance. But the memory of undoing a spark plug for the first time has stuck with me.

I have never had any formal education in motor mechanics, but I have spent endless hours learning it for myself. I have had various project cars and motorcycles, and kept a close circle of like-minded friends.

I read books on the fundamentals of motor vehicles to learn how carburetors and distributors worked. I toiled over internet pages attempting to understand gearboxes and differentials. I brushed up on how the suspension geometry affects the handling of a car. And I bought an extremely unreliable motorcycle which kept me busy out of necessity.

I'm not claiming that you must spend every waking hour focused on achieving mastery skills in the inner workings of brake calipers, but you should have a fundamental knowledge of the systems you will encounter on a car (or bike, boat, plane etc.). If you wait for a formal education (see the next section) you will be at a disadvantage for two reasons:

1) Your formal education may not provide the hands-on experience you are expecting. This means your knowledge is purely theoretical as you have not yet seen it in action.

2) You will be behind the majority of your peers with respect to knowledge and understanding. You will be playing catch-up whilst still trying to take on the new material.

At the end of this book is a list of recommended reading if you are essentially starting from nothing.

An often overlooked fundamental aspect is a knowledge of what is happening within the industry. This doesn't mean knowing everyone, in every team, on every grid, in every championship. What it means is, watch some races. Take an interest in Formula One, NASCAR or the World Rally Championship. Get yourself into a position where you have a valid opinion on a race series so that you can hold an intellectual conversation on it. When meeting members of the industry, being able to talk about what is going on shows you have a real interest and certainly won't do you any harm.

Formal Education

One option is to take it upon yourself to learn what you need to know. There are untold amounts of courses available online (with varying levels of credibility) that can teach you the very basics. However, if you are serious about making it, you will in all likelihood, need a degree or an apprenticeship.

Degrees

All around the world there are respected establishments offering relevant degrees for motorsport. This doesn't mean you have to seek out a specific motorsport degree however. Whilst I recommend something specific based on personal experience, I have had plenty of friends, team mates and colleagues in the past with less focused disciplines. A degree in Mechanical Engineering will not put you at any disadvantage to someone with a degree in Automotive or Motorsport Engineering. Similarly, a degree in Aeronautical Engineering will see you thrive in an aerodynamics arena, discussed in detail later. I would stress that all of these options have the word "engineering" within them. As stated previously, for maximum exposure to the frontline of motorsport, engineering is where you need to be.

Whilst studying, take every opportunity that presents itself to you. This statement may seem general or open, but it will become clearer as your studies progress. Engineering degrees are exceptionally social environments due to the inherent nature of those who undertake the course. There are always clubs and societies that will further your studies. These extra-curricular activities are absolutely critical to securing your first job in motorsport. Your experience gained from doing something *more* than your degree will set you apart from the hundreds of candidates vying for the same job. Below is a list of clubs I myself participated in whilst studying:

- The Student Motor Club – Like-minded students who enjoyed a more "hands-on" approach to car ownership. I was actually chairman of this club for 2 years.

- RC Car Club – Remote control cars might not sound exciting to everybody, but they come with the ability to fully setup suspension geometry. An excellent way to move from a textbook to the real world on an extremely low budget. Some even allow for aerodynamic optimization as well!

- Formula SAE – 100% essential. So essential in fact, that it has its own section below:

Formula SAE

Formula SAE, (sometimes referred to as "Formula Student") for those unfamiliar, is a project undertaken by engineering students all over the globe. A single seater race car is designed and built entirely by the student team at each institution that enters. The competition has grown from just a handful of entries, to hundreds of entries at events all over the world. The competition is split into two classes detailed below:

Class 1 – This is the real deal. From scratch, design and manufacture a single seat race car that fully adheres to the rules of competition. Each car is scrutinized prior to competing to ensure compliance, before being driven (by students) in several dynamic events – an acceleration run, a skid pad test, a sprint race and an endurance race.

The team must also participate in "static events". This involves giving a business presentation on the car, explaining the specific costs of certain areas and showing judges a complete business plan for mass manufacture of the car. Also included in the competition is a technical review where students must explain different areas of the car, showing how they have been innovative and creative in the design and manufacture.

Class 2 – This is everything as above, but without ever actually building the car. There are no dynamic events in Class 2 (as there is no car), but all of the static events are completed. Whilst this may sound like a lot of work for no return, I guarantee the return from Class 2 will be priceless when it comes to Class 1. Think of Class 2 as a low risk Class 1. You can try out ideas and designs and get most of the class 1 work done ahead of time!

It really cannot be stressed enough just how crucial Formula SAE is to getting a job in motorsport. When applying for jobs after university, my role and achievements within Formula SAE were more valuable than my degree itself. I honestly believe that my degree classification had less baring on job offers then Formula SAE did.

Formula SAE is not just a résumé filler. It taught me more than the rest of my degree combined, and I am not saying that just to sound poetic. It is the absolute truth. Here are the five top things I learnt from Formula SAE:

- Part Design – design for manufacture / design for installation. Although mentioned at points throughout my degree, it was not until I applied this that I truly learnt the importance of these principles. A part can be designed beautifully, and function flawlessly on a computer screen. If, however, it cannot physically be manufactured, then it is nothing more than a pipedream. Equally critical is the installation of that part. Designing "special tools" to install your new wonderful part is not time well spent!

- Team Member Reliance – I will cover working in a team fully later in this book, but I want to focus on a specific aspect I learnt. That is reliance. You will quickly learn who in a team can be relied upon to complete their work on time and to sufficient standard. You don't want to be the person who doesn't! Waiting on other people to finish, or worse, re-doing their work due to poor quality is time consuming and extremely inefficient. A core team of about 10 to

12 people was all that was really required for our Class 1 car. The team of 60 people was hopelessly inefficient. Fortunately, there was a high dropout rate, meaning that by the last few weeks were seamless.

- Lead Times – When ordering parts, they won't always arrive the next day. They might not even arrive the next month. I learnt how critical it was to be aware of lead times for parts and to factor that into the manufacturing process. Lead times can quickly start to stack up and derail a project. Stay on top of them!

- Commitment – Formula SAE is where I really learnt the meaning of commitment. Admittedly, throughout my degree, there were a few times where I stayed up all night to finish a piece of work or a project. But I would always sleep for a day afterwards. With Formula SAE it was different. Formula SAE opened up extra hours in the day that as a student I was barely aware of. We would start work from as early as 6am and get a few hours in before lectures started. Then, in the evenings, we would keep going up until midnight. Weekends were also consumed, with 15 hour days not infrequent. If you cannot commit to Formula SAE, you won't make it in motorsport.

- Politics – Perhaps the least known about aspect of motorsport, but definitely something you need to be aware of. It's not all just about what happens on the track that makes a good team. The atmosphere within a team is critical to its success. Any issues between team members (or faculty members for that matter) were resolved swiftly and politically. There was no favoritism, there was just objective action to overcome the problem. Learn to fight for the greater good and know when to back down. If you are always at loggerheads with others, not only will you drag the team down, but you won't get asked back!

Formula SAE is an unspoken requirement for any graduate job in motorsport. Adrian Newey, arguably one of the greatest Formula One engineers in history, has stated before that he simply will not recruit an engineer who has not gone through the Formula SAE program. The variation in powertrains available within the competition also opens the scope of your learning and understanding to new levels. You can run a standard gasoline engine, or a slightly more adventurous E85 one. There are fully electric powertrains (which currently are untouchable with regards to outright acceleration) or even hybrids. Formula SAE provides exposure to these possibilities and allows you to dig as deep as you like.

Formula SAE is by and large the biggest and most popular competitive motorsport arena for university students, however there are multiple other SAE administered competitions such as Formula Hybrid, Mini Baja and even a ¼ scale tractor competition. All of these will provide the same lessons and same exposure so pick the one that appeals most to you.

Feel free to claim the lessons learnt above for yourself; these are after all things you really do need to be aware of. But my intention of this chapter (indeed this book) is not to provide a list of buzzwords. My intention is to provide a framework on which to build your own knowledge.

So, one last time – DO FORMULA SAE!

Apprenticeships

Apprenticeships within motorsport are actually fairly commonplace, and there are numerous sources where you can see them advertised, particularly online.

Depending on your goals and timescales, you can opt to apply for the more competitive "Number 3 Mechanic" style apprenticeship, or the less sought after, but on equal standing, "Manufacturing" apprenticeship.

The Number 3 Mechanic will get almost immediate track side exposure and hands-on experience working on the car. You would of course be supervised at all times, but quite quickly you will be given specific responsibilities on and around the car. You will undoubtedly be given the less glamourous jobs (I know one apprentice who was tasked with cleaning a dead bird out of an engine bay after a rally stage), but you will get a priceless experience. You will soon find yourself intrinsic to team, attending every race and test day, as well as working full time back at base between events.

On the other hand, as a manufacturing apprentice, you will unfortunately have less exposure to running a car. You will however, be exposed to the brains behind the race operation. You will meet with design engineers and understand concepts before creating the parts to go on the car. Most teams will get you *some* access to a running car at points throughout an apprenticeship, but it is by no means guaranteed.

It's easy to see why the "Number 3 Mechanic" apprenticeship is the more sought after. However, completing either of the above options will grant you access to the industry. That access can be used as a stepping stone towards your ultimate goal. Don't rule out staying at base for the start of your career and learning what is happening behind the scenes. When your trackside time comes, you will have a thorough understanding of what has gone in to creating the car. This makes you astutely aware of the effort and money involved, and so will in all likelihood cause you to work that much harder!

Coupled with the apprenticeship roles, you may be able to get work at race schools, especially over the summer months. This will provide an excellent hands-on mechanic experience in a real motorsport environment. Race schools are more common than you might believe so it is worth investigating your local ones!

Specializing

With the exception of a few select engineers, there are no jack-of-all-trades in motorsport. Specializing in an area of interest to yourself is actually more sought after than someone with a general and basic understanding of multiple systems.

Specializing will set you apart during any recruitment process and will allow you to target a job that really appeals to you. Teams will never be recruiting for someone who knows everything there is to know about their car. What they are looking for is a fundamental understanding of most parts, and an acute understand of one or two. They are recruiting to give themselves a competitive edge, and so they want someone who can add to the knowledge pool within the team. As someone with no specialism, you won't be contributing to that pool, and therefore you are not desirable.

Some specific areas are listed below. I have tried to cover the "back stage" areas such as design and development, as well as the track side opportunities as well. Consider which environment is best suited to you:

Aerodynamics

Aero, as it is referred to, is seen by many as a bit of a black art. It is the engineering discipline associated with the airflow over an object as it moves, with the intention of reducing air resistance and increasing downforce (two frequently competing objectives). It is now the single largest area of research and development within many motorsport disciplines. On a Formula 1 car, aero is king. Every external element on the car has its design tweaked and compromised by aerodynamicists before being signed off to race. It can be an extremely fast paced world of computer simulation, wind-tunnel testing and rapid prototyping, often with new parts delivered to a car on the other side of the world within days of initial concept. Opportunities within aerodynamic engineering teams are increasing rapidly, but the people who get the jobs tend to be very good at what they do.

Superb knowledge of fluid dynamic principles will be critical to get as far as an interview, and comfort on simulation software will also be essential.

After design and manufacture, aerodynamicists fine tune the car during an event. They will be running various simulations at the track attempting to find every advantage possible in downforce and speed. If you are 1mph faster by the end of the straight or at an apex, you've done a good job!

Aerodynamic Engineers are highly sought after and highly respected, and it is definitely an area that continues to expand in pretty much every championship and series.

Powertrain

Ah. The noisy bit. The area that arguably attracts most people to motorsport. It certainly had a major influence on me.

Engine development is a hot topic at the moment with efficiency and fuel savings becoming key factors in winning races.

In recent years, engines have seen dramatic changes in displacement and power output, with turbocharger technology becoming a staple of many world championships.

You could opt for engine design, which is an extremely complex undertaking. I don't say that to put anyone off doing it, I have friends who are very talented at it and are doing very well for themselves. But this can overlap with the aerodynamics mentioned above. Rather than worrying about airflow over a moving object, you concern yourself with moving air through an enclosed system.

In an engine, air is power. Get more air in, you'll get more power out. So once again fluid dynamics and computer simulations will play a central role for a designer.

To get more track exposure, an engine calibration engineer could be a calling to consider. Depending on the championship and the team, there may be only one engineer in charge of an engine, or there may be half a dozen people making sure its running perfectly.

Calibrating an engine is again often considered a black art, but it can be learnt. It is something that requires experience and cannot be cemented from reading a book. You may be required to fine tune the boost control for a varying atmospheric pressure. Or find a sweet spot for an individual driver when it comes to traction control. These are not simple tasks, and the intricate relationships between different engine systems must always be considered.

Once again, you are looking for small gains, but it is the accumulation of subtle improvements that sum to make a competitive team.

Something that is fast becoming a very significant aspect of powertrain development is electronic powertrains. Electric motors now have their own championships (Formula E) and are definitely going to be a future technology that will continue to develop.

Formula E is already an entirely electric series that is encouraging battery and motor technology development, the likes of which are already making their way into road cars.

Chassis, Brakes and Suspension

The way a car "feels" to a driver is impossible to really quantify. Every driver will have a different preference when it comes to the balance of a car, and you simply will never please everybody. Under this heading I include all of the inputs and feedback the driver gives and receives, with the exception of the engine. It is what a driver feels through their hands in the steering, through the brake pedal, through the seat and even through the harness keeping them there.

Initial design of a chassis, braking system and the suspension components will make or break a championship. It can mean the difference between an exquisitely handling car that inspires confidence and great feedback for a driver, or an unforgiving, un-drivable pig, that scares your driver into slowing down.

As a suspension or chassis engineer, it is your responsibility to fine tune the setup of a car to best reflect how the driver wants the car to feel. Bearing in mind that a driver's preferred "feel" may not technically be the fastest setup. You will be responsible for deciphering logged data and objectifying the balance of the chassis as a whole. Is it understeering? Is it oversteering? Is it rolling too much? Or not enough? Are wheels leaving the ground? And if so, is that a problem?

Once again you will be faced with a labyrinth of interconnected settings, each affecting the others. Ultimately, your goal is finding the best compromise between keeping the chassis balanced and stable, and keeping the driver happy. Carroll Smith felt strongly that a mechanic should have at least attended a driving school to better understand the behavior of a chassis. This means you might get the opportunity to learn how to drive quickly as part of your job. And who doesn't want that?!

Electronics

If your intentions do not lie in classic car racing, then chances are you'll find yourself swamped in electronics. A modern racing car will have a significant amount of its budget and weight allocated to all manner of electronic wizardry.

Electronics are often seen as "black boxes" where only the chosen few know how to keep them working at their best. And it is true (as it is for everything) that experience is needed to really know what you're doing.

Under electronics here, I will include several areas. Engine Control Units (ECUs), Gearbox Control Units (GCUs), data loggers, dash boards, steering wheels and wiring looms will take up the vast majority of an electronics engineers' time. The overlaps here are obvious – the ECU needs to be calibrated by the engine calibration engineers for instance. The data from the data loggers needs to be seen by the chassis engineers. But making sure that those systems work together and communicate faultlessly is the task of the electronics engineer.

The system will start with a basic layout of components; what do we need and where can it be packaged in the car? From there, a wiring loom is designed to ensure that information and power can get to and from where it is needed. Communication to the system itself from a laptop or garage umbilical is included here also.

Depending on the team and championship, the track work of an electronics engineer will vary. Sometimes you will be there simply to fire fight in the event of something going wrong. Sometimes, you will be actively tweaking some systems to better suit the needs of other systems as the event unfolds. There will likely be an element of manufacturing as well, with loom repair and modification required at the track. Basic wiring, connector and soldering skills will really help here.

Electronics is an area that will continue to grow in complexity, and arguably one that is currently understaffed. Definitely an area with a future for those looking to specialize in the long game.

Hybrid Technology

Quite possibly the most public development for the motorsport world, hybrid technology has arrived and is most definitely here to stay. More and more championships are either de-regulating to *allow* hybrid technology, or re-regulating to *mandate* it.

It is a rapidly expanding area, with a deficit of knowledgeable engineers, and enormous scope for expansion.

Formula 1 and the World Endurance Championship are early adopters of hybrid and electronic technology and it will become a mainstream of motorsport within the next decade.

As an engineer working on these systems, you will be at the forefront of research and development, and budgets are only going to increase within this field. You will be at the pinnacle of the latest technology, and be able to watch your ideas move from paper, to the race track, and then eventually end up in road cars. This is technology that very quickly moves through the pipeline to OEM companies looking to sell cars based on fuel efficiency and environmental friendliness.

At the track, expect to be fine tuning the recovery and delivery systems to optimize the power output for the driver. Making the power delivery fast, but smooth enough to keep control is critical to inspiring confidence in pressing that button on the steering wheel. Equally, smooth regeneration under braking is crucial for corner entry stability.

Mastering this technology will earn you the respect of peers, and longevity within the industry.

Data Engineering

I include data engineering as its own section simply because it is the area from which I am coming to you. I have been a data engineer for teams in different championships, in different countries, and on different racing platforms, and cannot stress enough the importance of data in the modern racing environment.

This engineer will first download the data, and then analyze it to check systems are working as intended. If everything is in order (which is rarely the case!) then the engineer will move on to looking for areas of improvement. These areas will be discussed with the race engineers to decide a course of action. Maybe a change to an engine map is needed to allow a faster throttle response. Or maybe a suspension change is required to keep the inside rear wheel on the ground under cornering. These are things that may be covered by their own engineers, but as a data engineer in a smaller team, you will have exposure to all of these systems.

Although in some teams, each member of the above sections will handle their own data, it is often the case for smaller teams that a single data engineer will be present to help them all out. I recommend data engineering from a personal standpoint due to the varied and all-encompassing nature of the work. You will touch on every area on the car, dealing with strategy and setup decisions, and will delve into all areas of the car that are wired into sensors.

As mentioned in the Electronics section, everything on a modern racing car has wires coming out of it. Everything is monitored and controlled for optimum performance, and as a data engineer you would have exposure to all of those systems.

Summarizing a data engineer's role can be tricky, as it is just so varied, but there was a quote from a manager I once had that does a good job:

"Our role is to collect data and turn it into actionable information."
– Kirsty Andrew

That quote really sums out the role quite well. You can have all the data in the world, but it has no value if it cannot be used to gain an advantage. The ability to convert raw data into something useful has its place in almost every industry imaginable – in motorsport, it helps you to get on the podium.

Fabricators

Having a capable fabricator at the track is essential for a team to manage the unforeseen issues encountered throughout an event. Whether your driver contacts someone else or the wall, or a mechanical failure causes damage, the ability to repair and replace will be the difference between racing and not racing.

Fabrication, *good* fabrication that is, is a sought after skill. Traditionally, the role is held by older, jack-of-all-trades types, but the rate of retirement versus new skilled workers is tipped against the industry. New fabricators coming into the industry need a strong and varied set of skills. Welding, turning and milling need to be complimented with fiberglass repair, composite knowledge (covered in detail shortly) and even some wiring skills.

A fabricator working at the track will have a limited arsenal at his or her disposal. Most teams will have a welder (TIG is preferred, but often MIG is all that is taken to the track), as well as some basic hand tools. Depending on the level of motorsport at which you are competing, teams may have a small machine shop in one of the trucks. This will rarely compose of more than a lathe, a mill and a pillar drill, all of which will be likely be manual and not CNC, but the solutions I have seen created in these makeshift shops defies belief.

Fabrication skills take years to hone to the degree that you can be useful at a track, but it promises to be an exciting and varied line of work. Consider this option if you already have an apprenticeship behind you and a proven track record of work. You must be able to work very quickly, without sacrificing quality or accuracy. The solutions you create do not have to be aesthetically perfect; they simply need to be functional. They must survive the race. If everything goes to plan, you should have no work for the weekend. But this is motorsport, things rarely go according to plan. At least, they don't go according to *Plan-A*.

Composite Engineers

Composites engineers are finding more and more championships, and more and more applications for which they are required. Composites, primarily carbon fiber, made their debut in motorsport in Formula One in 1981 with the McLaren MP4/1 and quickly became adopted by every team. The success of carbon fiber was catalyzed by a devastating crash at the 1981 Italian Grand Prix, from which John Watson escaped completely unscathed. The combination of strength and low weight secured composites' future in motorsport.

A significant drawback with composites however is the difficulty in repairing and modifying parts. A small clash that results in body work damage can now no longer be repaired with a welder and piece of sheet metal. A composite engineer's job is to make sure the car runs, regardless of this damage. Patching is a difficult business, and without an autoclave at the track, you are restricted on the materials and processes available to you.

You might also be asked to manufacture new parts, or modify existing parts at the request of different engineers in the team. Tweaks to winglets, gurney flaps, spoilers and diffusers are often requested trackside. Without a Formula One style manufacturing department behind you that could fly out new parts to anywhere in the world within 24 hours, you must be able to develop and implement solutions without compromising reliability or safety.

Composite work can be attacked from either an apprenticeship or degree standpoint, and both can become very successful at their job. Your tasks may also include tweaks to molded driver's seats, adjustments to cockpit controls and creating mounts for new parts. Again, a varied role, reliant on an ability to work within timeframes and often under pressure.

Grid Support Role

Up until this point I have focused a great deal on being in a team. There is however another option to pursuing a motorsport career that does not tie you to a single team. That role is to be work for one of the many companies that support and supply the grid. Sometimes companies will send representatives purely to support their valued customers. Other times, their product will be mandated in the rules and their support engineers will be there to aid in the smooth running of the series or championship. Whatever the reason, grid support is an excellent opportunity when starting out in motorsport.

After university, I worked for Cosworth, and this was an exceptionally interesting and challenging role, with wide-ranging responsibilities and exposure to multiple motorsport disciplines. It was a fantastic opener to a motorsport career and one that put me in good stead for the following years.

Grid support can be labelled as many different things, some of the more common being "Applications Engineer" or "Customer Support Engineer", but are, at a high level, the same role. Your job is to support the application and use of the product (or products) that has (have) been supplied to teams at events.

Working for Cosworth, this involved supporting the electronics packages that they supplied. Sometimes this was mandated equipment for the championship (such as in the World Endurance Championship and British Touring Car Championship) and other times customers had chosen to use the Cosworth product and required expert support for just their team.

Working to support an entire championship is a huge responsibility, and you must have a lot of knowledge surrounding the product you are supporting. The teams will be indirectly paying you to be there, through championship entry fees, and so expect a level of expertise in excess of their own. You will be presented with problems that cannot be solved by common sense and the "usual fixes", problems that require an intricate understanding of the inner workings of the products.

Some areas that tend to be excellent for grid support roles are:

Tire Suppliers – Tire manufacturers will send an army of engineers to championships at which their brand is the sole supplier. You will have access to the cars after testing, qualifying and racing to analyze the tire behavior and, if necessary, advise the engineers on improvements. Are the tires working too hard? Are they overheating? Maybe they are suffering from uneven wear? Reading tires is a black art, so if you can do it well you will be highly sought after on the grid.

Engine Suppliers – A lot of championships will have a single or just a few engine manufacturers who supply homologated engines to the teams in the championships. This keeps the racing close, but means the teams themselves are not responsible for the engines. As such, engine suppliers will send engineers and mechanics to the events to support the engines and keep a close eye on the health of the units. These engine gurus will often become quite intrinsic to a team and can help find competitive edges, particularly with drivability.

Electronics – There are often datalogging systems on a racing car that are purely there for scrutineering purposes. The data in these systems is used after competitive running to ensure compliance with the rule book. As these systems are compulsory, engineers to *support* the systems are provided by the suppliers and championship organizers. Teams running their own standalone electronics will often opt for the same company's products to allow for easy integration between the systems, and to take advantage of the free support already on offer.

Brake Systems – Brake suppliers frequently make an appearance at the track due to the consumable nature of the product they sell. Pads and discs will be changed during the course of most race weekends, and having spares and upgrades available at the track is very useful for the teams. Coupled with this is the immense body of knowledge that the engineers and salespeople have regarding the product, opening up potential advantages and helping improve efficiency.

Fuel Systems – The fuel system on a race car is perhaps the most stringently mandated and safety critical aspects of the car. Fuel tanks generally will be "bladder tanks", and have complex pump, baffle and plumbing setups within them. A championship mandated fuel tank is not uncommon and something that means exposure to every team on the grid.

Transmissions – delivering the engine power to the wheels is tough work, and so the transmission of a racing car must be up to scratch. This, along with the changeability of gear ratios, means transmission engineers are a common site at the track. They will impart their expertise to race engineers who are looking to change their gear ratios to better suit their driver/track combination.

This list is by no means extensive, and there are countless other opportunities for working at race events without being embedded in a team. The above options, and those not listed, can also open the doors to being asked to *join* a team if your expertise, work ethic and knowledge are seen as a good fit. Grid support can be a tricky area to get in to, but it's fairly common to be headhunted back out of it!

The above areas are the primary sectors of engineering within a race team, and most readers should find one or two of them appealing to them more than the others. None are any easier than the other and none require less work. They all should be approached with respect and all contribute to a successful race team.

Professional Bodies

Professional bodies can be a real asset to your credentials as an engineer when applying for jobs, particularly when applying to the higher ranks of motorsport.

Two of the largest professional bodies are the Institute of Mechanical Engineers (IMechE) and the Society of Automotive Engineers (SAE).

Both bodies allow members access to vast libraries of text books, journals, magazines and periodicals, all of which can be immensely valuable during your study years, as well as when undertaking research professionally. Both societies host events all over the world to promote and encourage professional development, offering reduced price access for students and recent graduates. Both organizations assist with the Formula SAE competition, with the IMechE as the official governing body for the event in the UK, and the SAE administering the competition in the USA.

The IMechE is an excellent body for becoming an established and respected engineer, offering direct routes to chartership through universally accepted training programs. These training programs include proving your development as an engineer through research and furthering the body of knowledge within a certain area of interest. You will be required to document this work and have it assessed by adjudicators within the organization.

Professional bodies show an investment in your own education and career, and an interest in developing yourself professionally. They provide an excellent way of demonstrating these traits to potential employers as well as colleagues, and the benefits of membership are an excellent asset to have.

Personal Fitness

Personal fitness really wasn't something I had ever considered when entering motorsport, and I would hazard a guess it seems an odd thing to include to a lot of readers. It is, however, becoming more and more crucial to not only surviving in motorsport, but thriving.

Don't worry – you don't need to be superhuman to do this job. Unless you plan on making it in top flight championships as a driver, you can make do with a basic level of fitness. They key is endurance.

Motorsport doesn't run on fuel; it runs on caffeine. And caffeine has drawbacks. It makes you jittery, wired, and eventually it wears off leaving you vulnerable to mistakes. Whether you get it from coffee, tea, energy drinks or medication, beware of the come-down later in the day. My time in the industry has taught me that a more natural ability to maintain alertness comes from simply being fit and healthy.

Disclaimer: I'm not a doctor, nor a health expert, so seek medical advice before undertaking any exercise regime.

Now that the legal jargon is out of the way, go for a jog. A couple of miles a couple of times a week is all that is needed. The improvement in cardiovascular health means you will feel more alert and have the ability to work for longer.

The longer you can go without resorting to caffeine the better the engineer you will find yourself becoming. A reliance on artificial stimulants can be the downfall of an engineer in motorsport.

Don't get me wrong – a cup of coffee is absolutely fine, and it has its time and place. By the end of a 42-hour stint at Le Mans, I was chain drinking espressos. But if you can get through *most* days without it, you are doing well. And I promise, exercise will help with that.

Aside from alertness, the job can by physically demanding. You will often find that things need to happen quickly, and that means *moving* quickly to make them happen. Running between pit walls and garages multiple times during a race will wear you out. Holding a pit board at arm's length will fatigue you. Carrying tires, jump packs, tools and laptops will make you tired. If you can postpone these affects you will do better than those who can't. The ability to maintain your focus and energy for the duration of a race is an excellent strength to have.

I have also had the opportunity to run a lap of race tracks all over the world. This can become a social event with team members and even drivers. After the track is closed to traffic, you will see dozens of runners (and cyclists) hit the tarmac to boost their energy levels, improve their fitness and see the track. Be one of them!

Chapter Four

Landing Your First Job

"Choose a job you love, and you will never have to work a day in your life."

– Confucius

In this chapter, I aim to help you take those very first steps into getting some work. This chapter is by no means extensive, but it is an accumulation of advice and pointers that I have picked up from many people wiser than I. Not everything mentioned will be relevant to every job, but I have tried to cover the aspects which most regularly crop up, and most regularly catch people out. The contents here lay out some basic and fundamental steps you will need to take, as well as plenty of extras that will help you to stand out. Take heed my friends!

That Illusive Experience

Taking that first step will actually be one of the easiest parts of your journey into the motorsport world. Unfortunately, the reason it is easy is that you will have to work for free. Everyone I come across started out somewhere sweeping a garage or cleaning a car as a volunteer. And I do mean EVERYONE. The trickiest part with getting into motorsport is having work experience. Valuable experience is hard to come-by and vacancies are quickly filled so you must be prepared to help out just for the love it. This has a few effects:

- Potential Employers or Teams can see first-hand how much you want to work in the industry.

- It doesn't tie you down in case something paid *does* come up.

- It's very unlikely that any race team would ever say "No" to an extra *free* pair of hands.

- Having some sort of history working within motorsport will be crucial to getting to the next step: paid work.

I implore anyone trying to crack the motorsport industry to seek out local teams at the amateur and semi-professional end of the spectrum, or local race schools. Offering your services will be your best chance at securing some experience. Some teams and schools will be full. Some will be desperate. Get your name and your face in a garage at the earliest opportunity and be prepared to work hard.

You will most likely not find a job that pays you to be there. It's possible you will even end up out of pocket due to expenses of travel and feeding yourself. But I urge you to stick at it. This is where the vast majority of people drop out and don't succeed. This process is a rite of passage, and you would be surprised the number of big names in motorsport who started out this way.

Those who are uncommitted and the glory seekers amongst the masses are weeded out in this stage very quickly. The hard graft, inexistent pay, and long hours can quickly obscure the enjoyable aspects of the job and only those with the drive and passion required will make it through. Think of this stage as boot camp for motorsport. A grueling, brutal necessity that will prepare you for the career that lies ahead.

Targeting amateur and semi-professional teams is beneficial for a number of reasons:

1) You are far more likely to actually get a role. These teams have a lot less money than the bigger teams, and so will be far more grateful for the extra help!

2) The experience you gain will actually be far more valuable. This might seem counter-intuitive, but smaller teams, with fewer engineers means greater exposure for each person (which includes you!). Within a smaller team you are far more likely to experience everything from pit stops to pack-up. This will allow you to decide on the area that really interests you, and make you far more valuable when it comes to finding job number two.

I'll finish up with a word of warning with work experience; don't get stuck in an unbeneficial position. If your hard work and hours of toil are spent simply cleaning wheels or sweeping garages, consider looking elsewhere. Your time needs to be spent on gaining experience, and even though you are working for free, you don't have to settle for something that is not benefitting you. Take an objective assessment of the team you are with and ask yourself if there is room for progression and to actually *learn*. If not, go and work for free somewhere else. As stated earlier, volunteering will always be easy to – make sure it is mutually beneficial however.

Cleaning wheels and sweeping garages may well be part and parcel of the role, but it should not be all you are doing. You are there to gain experience in motorsport. Always ask about additional roles within the team. Take note of what other junior members are doing and ask if maybe you could give it a try. Put the broom down and put yourself out there!

This work experience is absolutely key to getting your first paid job. You need to add value if someone is going to pay you to be there. Make sure you get that value early on. If you are getting *bored* at a race track, then something is very wrong.

Social Media

Before we go any further down the recruitment process, there needs to be a word of warning with regards to social media. It is now standard practice for almost every company to research candidates online prior to inviting for an interview. Whether you are applying to a small local race school, or a championship winning IndyCar team, you will in all likelihood be checked out online.

The usual suspects of Facebook and Twitter will certainly be checked, as will LinkedIn and any other social media sites you are present on. Be sure that the picture portrayed by these sites is a positive one! Your profiles will be scrutinized by recruitment teams for any signs of what could be deemed anti-social behavior. Be that a photograph, a status update or a tweet, these things will be used against you.

Also, don't think that changing privacy settings necessarily protects you. Larger corporations employ external companies to do this research for them, and they have ways of getting around privacy settings.

Your Résumé

Your résumé within motorsport is a strange document. Some people will tell you it is absolutely critical and a must-have. Others will tell you that they've never had a résumé and are at the top of their game. It divides people and can become a minefield, so I'll try and lay out things best I can.

I'll assume that a résumé is something you *do* want to have in your arsenal when applying for jobs. (Personally, I think it would be crazy not to have one ready if someone asks!)

Whether your résumé is laid out in reverse-chronological order, grouped into sections, has a fancy border or the paper is vanilla scented can all be discussed somewhere else. I assure you it won't be included here. Favored trends for résumé format change year in, year out so best to check with a recruitment officer at university or college for the latest fashion. I will however state the following pointers:

- Your résumé should reflect your work ethic. It should be structured sensibly and laid out neatly. Bullet points are great, and blocks of text should be kept to a minimum.

- A résumé should be *exactly* 2 sides of standard letter-sized paper. Never more, never less.

- If your résumé is too long, you can reduce the font and margins but only to an extent. Don't allow it to become crowded and messy. Double and triple check that everything you have included makes you seem valuable to a team.

- If your résumé is too short, find additional relevant experience from your personal life (hobbies, interests) or expand on the information you have already. Increasing font size and spacing has a habit of looking immature and should be avoided.

- Always start the résumé with a *very* brief summary (approximately 5 lines) of your experience. No employer wants to know how you have "always dreamed of working in motorsport" or that "motorsport has always been a passion" of yours. I promise you that the vast majority of résumé's they read will make similar statements like that do not add any value to the document.

- DO NOT SAY "Since I was young...". This is possibly the single most overused phrase on a résumé or covering letter. Be more articulate and eloquent.

- Every detail should be justified in being on the résumé. Stacking shelves in a local store or serving fast food should only be included if it adds value in a motorsport environment. If you can't find value, state the employment and dates, and that's it. You only have 2 sides of standard letter sized paper so allocate that space to something meaningful.

- Sell yourself! Rather than stating your *roles* in each employment, state what skills your learned, and how you contributed to success within the job and company. Express how you developed yourself and learned from the experiences that were made available to you.

- Always include a section on hobbies. This shows you are a well-rounded individual and can hold a conversation on different subject matters. Never attempt to deceive here as this definitely an area you will be questioned on during an interview!

- Have multiple people proof read your résumé. Before an engineer sees it, it is filtered by the Human Resources department. They will have a checklist and it is either pass or fail. Making your résumé readable to people inside and out of motorsport will pay dividends.

- Read it, re-read it and re-read it again. A single spelling error could spell disaster for recruitment. Larger companies will filter simply based on this as it shows a lack of care in your work and a lack of attention to detail. Imagine sending out 50 résumés only to discover a spelling error on line 3.

Online Testing

Online testing is a relatively new addition to the recruitment process in motorsport. It has only really become central within the last few years, but you can expect to encounter it whenever applying for roles within larger companies or teams.

The online tests will vary in format and structure, but they will assess your problem solving and deduction skills in abstract ways. For the most part, you will *not* be asked motorsport specific questions or tested on your engineering knowledge. The tests are designed to assess critical thinking and numerical reasoning, rather than specific academic knowledge. These psychometric tests are designed to deduce both personality traits as well as aptitude in relevant areas.

You will be presented with practice questions before starting, and the tests will be timed. Definitely get in some practice of these assessments prior to attempting one for real. There are countless free websites offering example tests that will be very similar to the real thing.

The Interview

So you've made it through the résumé filtering, passed the online assessments, and now you are faced with an interview.

Firstly, congratulations! Getting this far through the recruitment process in motorsport is an achievement unto itself. You have done well!

The interview will be your first interaction with an engineer in motorsport. And let me tell you – first impressions count. Be well presented, but don't depart from your personality. If you usually have a beard, keep it; but make sure it is well kept. If you usually wear make-up, wear it; but not to the extent that it becomes a talking point. Don't create an illusion of who you are when going into an interview as this will add pressure in maintaining your cover. You want to go in as comfortable and relaxed as possible, and the best way to do that is go in as *yourself*!

Take time before an interview to re-read this chapter and write down some notes. Remind yourself of keys points you want to get across, and have the evidence to back them up ready to go.

I would recommend bringing in a pen and paper to an interview and keeping notes throughout. A tablet could arguably be acceptable, but I find them much less personal. You don't want to become engrossed in your notes and detached from the interaction you are a part of.

Start your notes with the date, location and company written at the head of the paper. Do this before you go in. This will give the impression that you are organized and proactive – two valuable traits which are difficult to prove in most circumstances. I also highly recommend writing down the names of those interviewing you – you don't want to get to the end of the interview and be thanking "sorry, what was your name again?" for their time.

I keep on about the notes as they are important. As your interview progresses, you may have points you wish to readdress later on, or questions that you would like answers to. Make quick, short notes, questions you are asked and answers you give. Again though, you are there to speak – not to scribe.

It is likely that the interview will be split into three parts; academics, experience and personal.

Academics

You will be asked about not only your grades, but what you actually learned and enjoyed whilst studying. Be ready for the question of "Why motorsport?". Use this section to deliver not only how academically sound you are, but also the experiences you have gleaned from studying. Recruitment agencies refer to these as soft skills and include things like:

- Communication
- Organization
- Time Management
- Teamwork

Draw on real life examples from your studies and explaining how those experiences developed some of your soft skills. It's all well and good saying "I did some group work", but you need to elaborate on why that is beneficial to the company. What was your role? What problems did you face? How were they overcome? Linking experience to soft skills is easy in this section so use it to your advantage. You can answer questions they haven't asked if you're clever enough with your responses. This will give them a far better overview of you compared to the next person they are interviewing with the same set of questions.

You can also mention at this point any professional bodies or academic societies that you are a member of, as mentioned in the previous chapter. Mention how these associations have assisted your learning and how you envisage them helping in the future. The access to the libraries, resources, websites and events can all be used to show an investment in yourself, your future and your career.

Experience

Your experience will be a far more technical area of the interview. You will be grilled on how much you can actually apply from your studies into real life. Feel free to use your studies for experience examples (especially for graduate schemes, they will not be expecting you to have years of motorsport exposure behind you). When answering questions, give as much detail on the first pass as you comfortably know. Stay within your comfort zone but talk with authority. Don't try and lie – the chances are stacked against you that the senior engineer interviewing you is more informed than you are.

As you finish each answer, expect to be asked to elaborate, to expand further. This is a common technique used by interviewers to weed out liars, and those who will crack under pressure. Neither trait is going to make it in motorsport. Keep answering, qualify answers with "I'm not 100%, but from my understanding…" or similar. Use the knowledge you have to make educated guesses and don't be *too* afraid of getting something wrong. If you get to a point where you simply have no idea (and that is where they will try and get you to), definitely say "I'm sorry – I don't know". If you can think of a way that one might find the answer out, include that. It shows initiative. But I will repeat again – DO NOT LIE.

Personal

Usually towards the end of an interview you will be given the chance to talk about yourself. By this point you have likely covered your motorsport interests by calling on them for examples of experience. If there are any left, mention them now so long as they add value. State why they make you a great candidate for the role.

Ensure you do not only talk about motorsport. Make sure you can talk about another hobby or interest that you pursue. If you were a member of any societies or clubs at university, now is your chance to discuss it. Show your passion for life is greater than motorsport alone. If these are people that you are going to be working with, long hours, over weekends, they are going to want to know you can hold a normal conversation.

Try and use your hobbies to your advantage. If you participate in sport regularly (especially cardiovascular sport discussed previously in the Personal Fitness chapter), mention how it has improved your overall energy levels and alertness. Hit the gym? Say how much that has helped working on your car at home by increasing your strength. Interested in computers? Maybe that has really helped your analytic and logic skills. It's easier than you think to relate seemingly arbitrary hobbies and interests to valuable skills for a potential employer. Take the time beforehand to plan this out. I stress again, don't lie! Getting caught out will guarantee you won't get the job!

Top Tip – Research the company before you go to the interview. Dropping in some key facts on their successes, current products and future developments that are available in the media will definitely help you out!

Phone Interviews

On a side note, phone interviews are getting more and more commonplace prior to a face-to-face one. Phone interviews allow a potential employer to remove a lot of the tricks mentioned above from your arsenal and decipher the essence of who you are, what you know and how you may benefit the team or company.

You should do your best to follow the above structure and pointers, taking notes will still be very valuable, despite the interviewer not knowing you are doing so.

The call will be scheduled for a set time and date, and the company will call you. Ensure you are somewhere quiet where you cannot be interrupted prior to the interview starting. Make yourself comfortable so you are not moving and fidgeting whilst the call is ongoing. Also, if using a cellphone, make sure you have a full battery and good cell reception – you don't want the flow of the interview to be interrupted by intermittent connections.

Preparing yourself for a phone interview should still involve proper research. You should also take a few minutes beforehand to calm yourself down. Your only means of portraying confidence is your voice so some calming exercises will help remove any shakes in your vocal cords.

Assessment Centers

Assessment centers are daunting places. You are pitched against other candidates for the same job as you but you are tasked with working together as a team. It is an unpleasant structure of trying to be a team player whilst still standing out in a crowd.

Assessment centers are not used by every team and every company, but larger companies are much more likely to have this as a part of their recruitment process. The assessment usually takes place on the company's premises and is standardized for many job roles in the company.

A likely task you will be assigned is to give an individual presentation on a pre-selected subject, or may be chosen by you. Often this presentation can be prepared ahead of time, before attending the assessment. Remember that audience will not necessarily be engineers, and you will want to stand out amongst your competition.

You will certainly be given a range of tasks that involve showing your ability to work as a part of team. You will be under constant scrutiny about how you communicate with team members, the role you undertake and what your contribution to the outcome was. Remain courteous with your teammates as well as the observers, and don't come over as overly competitive. You won't necessarily know how many positions are available, and you will most likely meet some of them again on your first day… assuming you get the job.

And that's pretty much it. That's the best guide I can come up with to help you secure a job in motorsport. There will no doubt be additional things that I have not had experience of, and therefore can't inform you about, but the previous pages should certainly give you a head start against almost all of your competitors for the job you want.

Visit www.startingonpole.com and share any experiences you think would benefit others. You can also keep up to date with what is new in recruitment within the industry in the newsletter and webinars.

Chapter Five

THE NITTY GRITTY

"Quality means doing it right when no one is looking."
— Henry Ford

It has taken a while to get there, but here we are; at the track. The track environment will be a huge culture shock to anyone who is starting out. Whilst you may be lucky enough to encounter perfect sunshine and all the glitz and glamour seen during the Monaco Grand Prix weekend, the chances are your first experiences will be somewhat removed from that ideal. It will probably be a test day, in the middle of January at a cold and wet race track that was cheap to hire. But, hey-ho, you made it! Let's go racing!

The diagram below outlines some key areas to success within the industry.

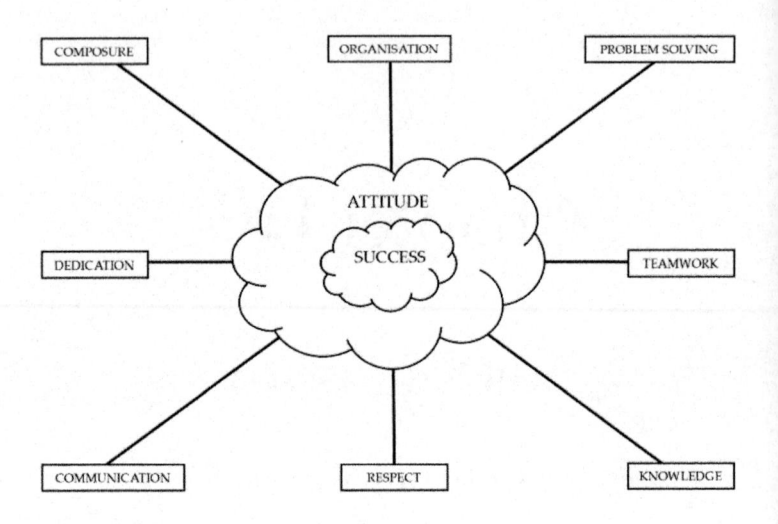

```
COMPOSURE      ORGANISATION      PROBLEM SOLVING

              ATTITUDE
DEDICATION    SUCCESS            TEAMWORK

COMMUNICATION  RESPECT           KNOWLEDGE
```

Attitude

Everyone I spoke to about writing this book mentioned how important the *attitude* of new engineers is. When you arrive in any team in any role, you will first and foremost be judged on your attitude.

"I find attitude one of the biggest assets in any new people coming into the industry. I know a "can-do attitude" is a cliché, but it describes it well!"
— Gummi Gudmundsson

Gummi is a good friend of mine and someone who has experienced numerous new, young and eager engineers attempt to tackle the industry. He has witnessed the spectrum of attitudes that will appear in all walks of life, and knows almost instantly who will and who won't make it in motorsport.

Your attitude is obvious from day one. Whether you are the quiet and reserved type, or the outspoken and outgoing type, your attitude will define the trajectory of your career. Being a genius might help, but having a positive, unfaltering attitude will carry you further than you can imagine when starting out. The ability to persevere against problems will set you apart from peers who give up at the first sign of trouble. You won't ever be everyone's friend, but getting the job done, no matter what, is a trait that will get you noticed.

The diagram on the previous page outlines the importance of attitude towards your work. It acts as a shield or a filter to success. You can have any number of positive traits about yourself; knowledge, organization, communication, but without the right attitude you will not succeed.

There will always be a select few who make it through with what could be deemed the "wrong" attitude, but they are few and far between. You will encounter these people during your career, and they are covered in the next chapter a bit more, but they achieve success through being the absolute best at what they do. As a new engineer, you cannot work that way. Your attitude, a willingness to help and a willingness to learn, coupled with positivity and perseverance will see you through almost any situation that crops up.

Efficiency

One of the first things that will likely strike you at the track is the efficiency (or lack thereof) within the team. Wherever you work in the world, and whatever championship you end up in, the more efficient teams will be at the front of the grid come race day. Your number one goal during the event is to be the most efficient person in the most efficient team.

Inefficient teams tend to be disorganized and caught off-guard. They do not adapt well to last minute changes such as rain, penalties or delays and as a result their race suffers.

Being efficient is a very difficult thing to teach. Efficiency is not simply completing a task in the shortest amount of time; this is often an error of inexperienced engineers at the track. It is much more about getting the best results for the minimum effort.

Minimum effort does not equal laziness. What it does is frees you up in a shorter time to start on the next task. Thereby, you as a single person at the track will tick off more jobs on the to-do list than others. So long as those tasks are completed well, you are being efficient.

Each team will strive to achieve the same things and complete the same tasks, but they will each have individual routines and processes for doing so. These processes will have been honed over months and years and should be a good representation of efficiency. Relating back to That Illusive Experience, the more teams you have worked for, the better grasp you will have of this. You will undoubtedly see areas of improvement between teams (no team is perfect), and with time and experience help to pass these efficiencies on to the new teams you become involved with.

One trouble you will experience during an event is that, due to the high pace of the work, you won't get the opportunity to try new processes as often as you would like. A race weekend is a ritual of rehearsed procedures, proven to work and proven to be low risk.

Suggesting new and untested ideas can frequently be met with skepticism and pessimism. An "if it ain't broke, don't fix it" attitude is actually a solid outlook for a lot of motorsport situations.

The solution? Write it down and try it out later!

Keeping notes is discussed in greater detail later (see Organization), but you will find teams and managers far more compliant and open to trying new things during down time between events, or even at test days. If you spot an area for improvement and can prove yourself right, you will quickly earn the respect and gratitude of those around you.

Communication

Effective communication is such a fundamental skill within motorsport that I toyed with the idea of giving it an entire chapter of its own. Being able to discuss issues and solutions on various different levels will play a key role in the impression you leave on people.

Many issues you will be required to solve as a motorsport engineer are complex and compound. Multiple systems integrating with each other, moving parts, high speed engine, chassis and aero dynamic variables, and a complicated array of connections between everything mean that the answer is not always easy to figure out.

If you are the bright spark in the team, you are able to deduce what has gone wrong and, more importantly, *why* it has gone wrong, you need to be able to communicate that issue to a variety of people.

Engineers will want to know the intricate details of the issue, and they won't want to wait for a long-winded report. A concise summary, just a few sentences, needs to be enough for them to decide on a course of action.

Tailor your answer for the technical level to whom you are speaking with. The procurement and financial teams will not be overly concerned about exactly what the mode of failure was and the precise conditions leading up to that. They will want to know what failed and if it is likely to happen again. They want to know the impact on the team's coffers.

Drivers are trickier to handle and are discussed in more detail later, but different drivers will want different things. Some drivers will be on the same level as the engineers; having a deep understanding of the components that have failed, and will also be interested to know the likelihood of it happening again. They would also want to know the impact on their weekend if it were to happen mid race.

Being able to talk to these different levels will set you apart from a lot of the people coming up through motorsport. It makes you more valuable to the team and a better asset to keep around.

Talking on a radio is a skill unto itself. Communication via radio (be it handset or headset) should be kept to an absolute minimum. You will find a few people in each team who use the radio a lot to broadcast messages to everyone in the team (garage managers, team managers, logistics managers), as well as dedicated channels for engineers to talk to drivers. The chances are that, to start with, you will have "listen-only" access to drivers' channels. Pay attention to these channels as vital information is often fed back from the driver during events. You will get prior warning of punctures for instance, not waiting for the garage manager to broadcast it.

If talking on a radio, choose your words and phrases carefully. "No" can sound like "Go". "Tire" can sound like "Fire". If you can, choose words that cannot be misheard. It may sound strange saying "Negative" over the radio the first few times, but it can be a lifesaver (literally!). Qualify words and phrases if they need to be used so that meaning can be interpreted if the words are unclear:

Did he say "Engine Fire" or "Engine Tire"? Well "engine tire" makes very little sense so we can assume it's probably the former and therefore needs our attention!

Wording like this can also be necessary in a noisy garage, so don't just limit it to radio chatter. Be prepared to shout if needs be, but don't do it aggressively, and once you have been heard, pipe down so others can shout too. It seems strange, but mutual respect for shouting in a garage turns into a slick and effective communication strategy.

To finish on communication, write reports. After events do a formal report for your managers, or even just for yourself, on the comings and goings, what went wrong, what went well. If you tried something new, was it successful and can you quantify that success in either speed or time on the track? Writing technical reports provides a go-to log of the history of the car, and allows you and the team to refer back to previous experience. I have written reports for every event I have ever attended as a professional engineer, and it does become easier. You learn not to waffle and to be succinct in your writing. Include examples, photographs, data charts and tables if they are relevant and clearly show the point you are making. These reports could end up on the Team Owner's desk, so try not to be overly critical of individuals or team processes. Focus on telling the technical story of the race. Constructive criticism will always be welcomed by any team manager, but blatant moaning will not be appreciated.

Organization

Remaining calm and collected under pressure is the difference between many good engineers and the truly great ones. How do race engineers in Formula One keep their cool when multi-million dollar cars are heading around a track at 200mph, and their driver's red mist has come down? The answer is organization. Its knowing that whatever the outcome, they already have a plan in place. A proven sequence of tasks that resolves or rectifies the issue. They always have a Plan B and Plan C. Having that safety net takes the pressure off.

So how can you, starting out on your career, possibly have that safety net too? Well I'm sorry to say, you can't. That is something that can only be gained through experience. But that doesn't mean all is lost.

What you *can* do is to start making notes. This will make those first few threads in what will become your safety net and your pool of knowledge on which you can draw upon.

I have several notepads, each the size of my wallet, filled with notes, equations, sums and contacts. I keep a separate one for each team I support (or a single one for "one off" support events) and document everything that happens as the events goes along. This is similar to taking notes during lectures at university or college, only much more to the point, and without the daydreaming and doodling. Each note is never more than a line or two long (remember – it's the size of my wallet), and summarizes an issue, a change or an observation. The note does not have to include issue, root cause and solution all in one. It just a quick jot down of "I did this" or "I saw this". Even "I *thought* this" makes it in from time to time. These notebooks will become your books of knowledge. Many mechanics I know use a similar system and will have a notebook in their tool chests – torque settings, setup changes and faster procedures on jobs are all written down.

If you are one for modern technology, then an iPad or other tablet can be used for making notes also. I would point out however, that handwritten notes in a notepad tend to be faster, and a garage is often a dangerous place for what is effectively a thin pane of glass. Each to their own, but you've been warned! I've also never had a notebook run out of battery…

As a secondary source for my notes, I back up a large amount of them by hand to a cloud based note system. I use OneNote, although I have had plenty of colleagues who have had success with similar programs such as Evernote - chose whichever you prefer. Doing this after each day, or at the end of the event has the added benefit of reinforcing your musings from the day, allowing you to quickly recall that you wrote something similar should a problem arise again. Electronic copies of your notes are also always available. If something was written in your *old* notebook, you can still access it with an internet connection. This has proved priceless to me on numerous occasions!

Now that note taking is out of the way, we shall talk about your schedule. I highly recommend keeping a detailed calendar with a cloud backup. I use the Gmail calendar, but outlook works equally well. You want your phone, laptop and tablet to be in sync and to include all the details you require for each event. A cloud based calendar doesn't just set your dates; you can upload itineraries, hotel and track addresses, contacts and data files. And it will all be with you wherever you have an internet connection.

This synchronized life also prevents you from double booking yourself. You can split most cloud based calendars into business and personal, which means you won't book in a race weekend on your anniversary, or commit to two teams at the same time!

Setting times in your calendar will be personal preference. Some like to add time before and after an event to prevent overlap or overrunning of previous engagements. This works fine, so long as you have a standard amount of time. What can happen is the true start and finish times becomes hazy, and as such you may end up waiting around before an event a lot longer than intended. The other option is to put appointments in your calendar with precise start and finish times. You will need to remember to add your own buffer however. I use the latter method, but experiment and see what works for you.

Finally, be on time. There is a saying that floats around in motorsport:

> *"Five minutes early is on time."*
>
> -Unknown

If you need to be somewhere, be there early. Motorsport is fast paced on and off the track. Successful teams function like clockwork and you should endeavor not to be the cog that's broken. Punctuality will take you a long way in motorsport, and a lack of it will see you quickly demoted to non-essential and non-time critical roles within a team. The fast paced and ever-changing nature of the industry means that you must always allow contingency for the unexpected.

Knowing your Role

Each person within a team has a role. I apologize if that sounds like I am stating the obvious, but it has to be said. The point I'm making is not that people are simply doing their job, it is more that they are there for a reason. No one is at a track without a defined role to play within the team. It's crucial that you know yours!

Regardless of how much experience you have, even if it is your first day, you will have a position and place within the team. It costs money for each person to be there, and in some championships, team numbers are strictly limited. It is possible that, to start with at least, your role is one of observation, learning and being a runner (a runner is someone who does a lot of moving between people, conveying messages or getting parts/equipment to where they need to be).

Whatever your role is, embrace it. You will be the only one with your specific set of responsibilities. If you fail at those responsibilities, don't expect to be invited back. With more experience (be it real world or academic), you will be given additional roles to play in the team, to the point that the team will be dependent on your capabilities. You will get to the point where if *you* fail, the *team* fails. No one wants to be that person. Practice juggling and balancing your responsibilities whilst they are less critical. Put sugar in the wrong person's coffee? Not really a big deal. Underfuel a car before a race? That is a big deal!

Once you are in a team, try to quickly understand what everyone else's roles are. Learn who is who, and the hierarchy. Learn names and faces. This will make you much more effective at being a runner (if that's where you start), and also has the added benefit of spotting additional opportunities when others move on. Always be striving to do the next thing and you will be spotted as keen and enthusiastic. You will quickly progress up the ladder if you ask for a leg up!

One of the greatest things to experience in at the track is a slick pitstop. Depending on your role and the championship, you may be a member of the pitcrew as a primary or secondary role, and your abilities during those few seconds can mean the difference between a victory and loss. No pressure!

If you are not required at a pitstop, stay completely out of the way. Those teammates that are doing it will have a honed routine and will not want you standing anywhere that could get in their way.

Sometimes, during qualifying or practice, the pitstops may be less formal. Figure out if you are required to interact with the car when it comes in and be astutely aware of your surroundings. If tires are being changed, don't get in their way. If refueling is being done, ensure you follow whatever safety procedures are in place, be that wearing fireproofs or staying back in the garage for example. If you have time, take things nice and slow. If in doubt, stay away.

Simple rules that can keep you out of a lot of trouble at the beginning include:

1) If you need to touch the car, ask for permission from the mechanics. You won't necessarily know if they are draining a fuel system, charging a battery, or performing a setup change unless you ask.

2) If you borrow tools, always put them back where you got them from. A mechanic will spend tens of thousands of dollars on their tool kit over the course of their career and you won't want to be losing their equipment.

3) Don't be afraid to hold the car up *if you have to*, but make sure you are certain. If you need to hold the car back for whatever reason let the crew know how long you will need to fix

whatever the issue is. Always finish fixing that issue *before* that time is up!

Confidentiality

Motorsport is a strange industry for sharing. You will rarely find any team more than a season ahead or behind the latest technology or innovation due to the amount of knowledge movement between teams. At the end of each season, there is a huge migration of staff. New and better opportunities, a change of scenery, or sometimes cloudy histories mean many engineers will change to rival teams or different championships. With them, they bring the specific technical knowledge they had access to and so the cycle of innovation continues. This movement of intellectual property is legal, legitimate and accepted by teams throughout motorsport.

What is NOT accepted is confidential information getting out during the season, and from current team members. Losing the edge due to a leak from within a team is unforgivable and will demolish the most resilient of reputations. Even accidental leaks can end careers in an instant.

When you're at the track, you will become familiar with people from other teams. You may share hospitality, hotels and social events, as well as a close-knit work environment. You will make contacts and friends with people from all over the grid, and all over the world, and these people are excellent people to know. Share stories and experience with them by all means, but do not give away anything that might affect your team's advantage.

It's given various names (Paddock Talk, Pit Lane Gossip, Motor Mouthing), but anything you say can and will spread far quicker than you think. Keep this in mind when discussing anything at the track. It is easily overheard, and impossible to contain. You really don't want to be the person who spills the secrets, even if it is unintentional.

From personal experience, I would also like to highlight the dangers concerned with removable memory devices such as USB sticks, portable hard drives, and email.

I have witnessed instances of these devices being left out during race events; devices which contain extremely sensitive and confidential information on race tactics, new technology or logged data. Fortunately, the instances that I can recall were rectified before any damage was done, but you must remain so careful. I recommend keeping the use of removable memory devices to an absolute minimum, and always wiping the data once you have finished.

Email can also be dangerous. Double check the recipients, triple check the content, and beware of anyone reading over your shoulder. Using "Reply to All" is often unnecessary, and can lead to embarrassing and unfortunate situations that are so easily avoided.

The Competition are your Colleagues

One of the largest shocks you will have joining the motorsport industry will be the attitude towards other teams at the track. These teams will remain religiously "the competition", but they should not be considered "the enemy". The distinction between these two labels is subtle, but critical.

The other teams in the championship are there to make your job more interesting and more enjoyable, and will also push you to continue improving. The competition provides the drive to make sure that you are developing and striving to better yourself, and the team. This, coupled with the fact that you actually have someone to compete against, keeps the job interesting.

You will witness a huge amount of comradery and sharing between teams on the grid. Spare parts, use of equipment, and even knowledge are all shared as needed. The reasons for this are two-fold:

1) There will come a point where you would like the favor returned. You need spare parts or find yourself in a spot of trouble, you want to have other people around you that you can call on if needs be

2) No one wants to win by default. You are at the event to compete and enjoy that competition. Winning because another team failed to start or failed to finish is not really winning.

Helping other teams out also gets you known. You will gain a reputation of being approachable, friendly and helpful, and these are great qualities to be known for when it comes to applying for your next job!

Health and Safety

Motorsport is a dangerous environment, and not just for the drivers. You will see the sign below (or one similar) at race tracks all over the world.

Fireproofs – Depending on your role, the working environment and the championship regulations, you may be required to wear fireproof clothing whilst working. If you are required to, you are required to for a reason. It is almost certain that these fireproofs will be provided by the team and will not be an inconvenience or expense for you. You will most likely (and hopefully) go your entire career without ever "testing" these fireproofs, but should you need them, you'll be glad you are wearing them. The fireproofs also mean that should the worst happen; you are far better equipped at tackling a fire or rescuing someone else from danger.

There are a lot of flammable things surrounding a race car. The obvious fuel, oil and brake fluid are compounded nowadays by the flammable nature of carbon fiber. Fire can spread very quickly, and your priority should always be the safety of yourself and those around you. Cars can be rebuilt, so get the driver out and only attempt to tackle the fire if it is safe to do so. Note that the car and driver DO NOT have to be from your team! Famously, Christophe Bouchut's Lotus caught fire at Fuji in 2014 during a 6 hour WEC race, and that fire was extinguished by rival teams in the pitlane as they were closest.

Racing cars are required to have on-board fire suppressant systems, activated by a switch marked with a red triangle, usually in front of the driver. Remember, saving the car is the lowest priority, but if you can get to this switch safely, do so.

The Pitlane – A pitlane is a live arena. It is part of the race track, and should be treated as such. When working in the pitlane be aware of everyone and everything around you. Listen on your radio for cars coming in, and also listen out for the klaxon that sounds when a car enters. The speed limit in the pitlane is usually around 40mph (although this can vary depending on championship, track and country) and racing cars are not designed with pedestrian impact protection in mind. Getting hit means a very bad day for all involved so be extremely aware of what's going on around you. Also be wary of cars pulling into the garage.

Don't be afraid to shout "STOP" if you see a hazard or danger. Sometimes there will be VIPs or press in the garage area that are less familiar with the environment. Keep them out of the way as best you can.

The Pit Wall – The pit wall separates the race track itself from the pitlane. Mostly, you will be facing out towards the high speed track traffic, but you will have your back to any traffic in the pitlane. Again, be very conscious of yourself, and those around you. Usually, access to the pit wall is restricted during the race start, and you will frequently find a retaining fence to prevent debris from crashes from making its way into the pitlane. If manning a pit board, always watch the *approaching* traffic – you do not need to worry yourself with cars that have passed you. Be wary of leaning over the edge of the wall as you can become a casualty in the event of a collision on track. It is also not unheard of for cars to pass so close to the pit wall that they can clip the pit board; be wary of this as cars approach.

Long Hair – This should be self-explanatory and you will probably have heard this in whichever training route you have come through. If you have long hair (boys *and* girls) it MUST be tied back. You will be around running engines, moving cars and powerful tools and equipment such as pneumatic wheelguns. Fortunately, I have never seen it myself, but it doesn't take a huge amount of imagination to understand the damage that could be done by loose hair getting caught. The same is true of facial hair, gentlemen!

Hot Hazards – A race car that has just come in from the track will be seething with heat. The engine block will have surface temperatures of a couple of hundred degrees, the exhaust and brakes can operate at temperatures in excess of 1000 degrees (that's Celsius!) and even the tires can be 80+°C. Mechanics that are working on these areas will typically wear heat resistant gloves and forearm sleeves to protect themselves. Be aware of these hazards if working on the car freshly in from the track.

Rotating Hazards – As mentioned above, there are plenty of things that long hair can get caught in. But the same is true for loose clothing, lanyards and jewelry. Be aware of engines, drivetrains and any tools or equipment that spins – most of it is far stronger than you are and will not take any prisoners. Disconnect airlines and unplug power tools before changing connectors or drill bits.

Chapter Six

TEAMWORK

"Talent wins games, but teamwork and intelligence win championships."

– Michael Jordan

I have heard parallels being drawn between the comradery found in soldiers and teams within motorsport. Everyone in the team has a role and a purpose. Without that person, the team cannot function effectively, and that team will lose. Losing in a military setting are obviously far more severe, but the similarities are there.

On a side note, I am aware of several teams and organizations that use motorsport as part of a treatment and rehabilitation plan for veterans – check out local teams and charities for more information.

Thick Skin

One of the fundamentals of working in a team that is frequently under a lot of pressure is to develop thick skin. On top of the usual "having a bad day", you will come across many people who are highly strung and under immense scrutiny. There are a few people in each team upon whom winning and losing is a professional and personal responsibility. Often, these people will be veterans at their trade, and capable of keeping their cool under the most pressing of circumstances. However, there will always be those who can't cope.

It's quite likely that within every team you work there will be someone prone to exploding in anger for seemingly trivial reasons. And believe me, as the new guy, you'll will bear a lot of the brunt of it.

The fact of the matter is that scathing attacks and shouting matches will be part and parcel of working within some teams. To survive in this environment, you need to decipher what the shouting is all about. If you have genuinely done nothing wrong, then take it on the chin, put it down to them having a bad day, and move on. Maintaining your professionalism under circumstances like that not only looks great to peers and managers, but is superb practice for keeping calm when it really matters.

If, upon scrutinizing, you have in fact made a mistake, then accept it. Acknowledge to yourself that shouting and swearing is not the most constructive way of getting their point across, but you should endeavor to right the wrong. Fix your error and don't make that mistake again. Again, a superb opportunity for bettering yourself.

What you shouldn't do is shout back. Especially when starting out. You will come over as combative and insubordinate, and at an early stage of your career you will not have the reputation, respect or authority to back up what you're saying. Whether you are right or wrong becomes irrelevant as soon as you raise your voice unfortunately.

By all means raise concerns with others within the team. Speak with the aggressor once they have calmed down and use the opportunity to understand the situation. Never take it personally – it isn't. Never assume you will be fired – you won't. Chalk it up to experience, thicken your skin, and carry on with your job.

These people can be very difficult to work with, and you may wonder how and why they have survived in the industry. The fact is that they are probably very good at their job. They get results. Regardless of their methods, if they continuously contribute to winning races they can get away with having some less desirable character traits. At the start of your career, you cannot. If you can avoid this as your career progresses, you will always be much more popular on and off the track.

Managers

Management in motorsport is a huge topic, and one that has a library of books all to itself. My intention is not to teach you to *be* a manager, but to work with them.

There is a spectrum of manager that you will come across. Some will be firm and aggressive and, as mentioned above, you need thick skin and a calm demeanor to work with them. Others will be completely on your level and talk to you as they would an old friend. And of course, everything in between exists somewhere as well!

When you first join a team, take the time to learn who's who and the hierarchy. This is particular important at the track as it means you can inform the correct person of issues straight away without having to be passed from person to person.

Also, crucial information to know, is which of the managers are technical and which are not. There will be times where you need to discuss a deeply technical issue with a financial officer or procurement manager. These people will not necessarily understand the intricacies of what you need to convey. Tailor your report, pitch or brief to whomever you are talking to. Being able to convey a need or issue to someone completely detached from the system in which the issue exists is an excellent skill in itself. Being able to *teach* a system shows a great level of understanding and managers will know this. Talking to them on a level that they require will be remembered and can lead to additional opportunities.

When talking to technical managers (garage managers, race engineers, technical officers), use precise and accurate language. Always keep things succinct so as not to impede on more of their time then intended. You'll be surprised how quickly you can get across what you need to when you impose time limits on yourself.

Peers

By peers, I mean those in a team with whom you spend the majority of your time. You may not necessarily report to the same managers, but you are on an equal level. You slot in at the same place on the hierarchy.

Work hard at building and maintaining relationships with your peers. If you need something doing that does not fall under your remit, then it will be a peer who does that task for you. The easier you make it for them to say "Yes", the easier your life becomes.

Your peers will also become good friends over time. You will experience highs and lows in the same way and at the same time. They will understand when you are under pressure and vice versa.

In my experience, the peers generally have excellent relationships. They are comfortable complimenting and criticizing each other as necessary and a great sense of humor and laughter prevails through most teams.

Motorsport is unfortunately a very cliquey industry. If you're not in, you're out. It isn't difficult to make friends within a team so long as you have a positive attitude, a willingness to help out, you're easy to get along with, and most importantly, a good sense of humor! Don't be fake, don't start or spread rumors and don't make anyone else's job more difficult than it needs to be.

Following this structure will help you easily slip into a team. What's more, as these peers move around the industry, often moving up the hierarchy, they will remember you. Another key ingredient in securing your next job.

Subordinates

It's unlikely that you will have subordinates in the earliest stages of your career, but at some point it will happen. At some point you will be responsible for other team members. Their actions and attitude will reflect on you, so a large part of your management will involve nurturing them to become ideal "peers" as mentioned above. Like within management, there is a spectrum of subordinates.

You will have people under your management that get angry or frustrated at the slightest hiccup, and will need a fair amount of your time dedicated to them. Ensure you don't fail to address any concerns they have. It's all too easy to think they are negative and therefore of little or no value. Often, they will become excellent engineers with the correct guidance.

Others will be eager, determined and dedicated; driven by a desire to win and acknowledge that they are lucky to be in the position they are. These people tend to be easy to manage. That doesn't mean however, that you are able to neglect them!

A bad teacher is threatened by their students being smarter or more talented than them. A good teacher hopes that this is the case!

– John McCrory

John provided this quote to me, and it really resonated. It's definitely something you should keep in mind when you stop being the "new guy" and start training and helping new recruits.

With all your subordinates, be friendly and encouraging. Consistent positive feedback is always appreciated so long as it is sincere. Always congratulate on a job well done and always have a thorough debrief. Criticism is seldom well received, but if you ensure it is constructive and looking for improvements, you will gain the respect of those around you.

Drivers

Drivers are a funny bunch. Depending on the level of motorsport you are competing at, the type of driver you come across will vary. What's written here is by no means definitive, and there are always exceptions, but my intention is to prepare you for the possibilities.

Golden rule: REMEMBER you are NOT a driver. Driver's (and your team mates for that matter) do not want to hear how skilled you are behind the wheel. You won't be offered a chance to drive, and the odds are very much stacked that the person wearing the crash helmet is better than you. Talking about how a lack of funding means you can't race comes across as arrogant and is a story that everyone in the industry has heard many times before. Don't be that guy.

Keeping your driver happy is one of the most crucial parts to finding speed in the car. An unhappy driver means lap times will suffer.

Unhappy can mean a few different things, so I'll split it up:

Confidence in the Car

If your driver does not have confidence in the car, then he or she will not be able to drive that car to its limits. They will become so wary of *exceeding* those limits that they will do the natural thing and avoid the limits completely.

Different drivers will want different things from a car. Jenson Button famously wants a solid and predictable rear in his car. If the rear of the car is planted up to the point *he* decides to induce some slip, then he will be happy. How the front of the car behaves is second to the rear. Button tolerates and drives around a degree of understeer, so long as he has the confidence to use the throttle whenever he needs to. If the car is not setup in this way, you will see lap times falter. He loses confidence in how the car will behave in the corners, and his ability to accurately predict the grip available fades.

Ayrton Senna on the other hand preferred a much more "tail-happy" car with a planted front axle. His ability to push the rear tires to their limit without exceeding it meant he didn't need to rely as much on mechanical grip at the rear. The result is a car that danced around the front wheels at his command.

Two very skilled drivers. Two very different driving techniques requiring two very different setups. Listen to *your* driver and keep them happy. If they complain the oversteer is catching them out, dial it out. If they say they can't get the rear of the car to fall in-line, dial it back in. Keeping your driver confident in the setup of the car will be a very quick way to improve lap times. (Note: Sometimes *reducing* grip improves the balance. Consider all options when setting the car up for each driver.)

The Red Mist

During a race, things out of everyone's control will happen. Whether deliberate or accidental, your driver, your car and your team will be the victim of circumstance at some point. It is usually at this point that a driver will register their "displeasure" over the team radio.

This is a critical difference between drivers, at every level of the profession. How they react to a bad day.

Losing positions, damage to your car and the endangering of your life are not things to be taken lightly. But the very best drivers are able to accept and move on, without allowing the incident to cloud their judgement. Sometimes a calm and collected word from a race engineer or team principle might be needed to encourage the driver to keep their head in the game and not get angry and overly aggressive; not to seek retribution.

Keeping your driver happy once a race has started is a tricky business, and not something you will likely be required to do at the beginning of your career. It is something to be aware of however. Learn how to spot and deal with the different types of drivers early on so that it doesn't come as a shock later on.

Amateur Drivers

At the amateur level of the sport you will find a vast spectrum of skill levels and attitudes. There are those who think they're good, and those who *are* good.

Those who think they're good can be very difficult to work with. Whenever a race or event goes against them, they will seek to blame anyone and anything but themselves. Dealing with these drivers will become frustrating, and I would point you at the Thick Skin section at this point. Don't feel alone when you come across these drivers – everyone in the team will be aware of the attitude they have.

I don't endorse deceit, but am aware of times where little white lies can defuse a situation. If a driver complains of lack of engine power for instance, it has been known for teams to tell them that they discovered an issue when in fact there was none. The resulting placebo effect alleviates the pressure on the team and the driver so everyone is happy. The danger with this approach is the "boy who cried wolf" theory. When there is a real issue, it will go unnoticed or uninvestigated. I don't need to tell you the consequences of this can easily cost a race or worse.

Drivers who are naturally skilled can be arrogant about their abilities, but in my experience, they tend to be quite humble. They let their driving do the talking and have no need to brag about their abilities as their trophy cabinet keeps filling up. Skilled amateur drivers tend to be far more interested in the technical details of the car and the setup. They will be interested in changes made to the setup of the car and the affect those changes should have on the balance. They will endeavor to get to know the team members and are aware that without each member, they can't race. They consider themselves lucky to be in a position that they can enjoy their hobby to the extent that they do.

Professional Drivers

Professional drivers are an entirely different breed. They will most likely have raced for years and have honed their skills to an incredible degree. Regardless of the championship, to reach a level whereby you are *paid* to compete shows a foundational level of ability that is ahead of the vast majority of people on the planet.

There are a few things that set professional drivers apart from skilled amateurs:

1) They tend to be better. A simple one to start, but the fact is to progress in motorsport as a driver, you need to win as an amateur. Once you have gone pro, you then get a car and track time to improve your skills

2) The level to which they are in tune with the car is phenomenal. They will have a good

technical understanding of the car, but are also able to give on-the-fly feedback about changes that have been made. The debriefs after driving are technical and concise, and give the engineers a lot of information to work with in order to improve the car.

3) They are extremely competitive. They are there to win, not to come second. This will drive them to (sometimes) do crazy and stupid things. Their desire to win trumps all else.

4) They are good with Public Relations. The only way they can fund a career in motorsport is with sponsorship. And to get sponsorship, you need to be able to sell yourself. Sponsors look not only for talent, but for a fan base and the ability to sell themselves. Professionals will bring all this to the table.

Note that the above rules *tend* be the case, but you will find some naturally talented drivers who actually don't know what happens to their suspension geometry when they go around a corner or how pressing a brake pedal slows a car down. The statements above do outline a "best case" scenario, but beware that there are exceptions to every rule!

More often than not, dealing with a professional driver is a great experience. As an engineer, you will learn a huge amount about the car, about the behavior and balance, and strangely, about psychology. A good driver could be faster in slower car, if he has more confidence in it.

If you ever get the chance to work with a driver on this level, I cannot encourage you enough to take it. More often than not they are great people.

Remember, that like everyone, they have good and bad days though! They won't be in a great mood if they can't race for one reason or another.

An anecdote from supporting Gabriele Tarquini:

I was doing some fly-by-wire throttle fine tuning for Gabriele during a testing session for the WTCC. He came in from a 5 lap stint and said he felt the throttle was not closely related enough to his inputs, stating "When I put the pedal to 20%, it feels like the engine is at 50%".

As a data engineer, I looked at the data. Pedal Position was at 20%. Throttle position was at 50%. Gabriele had no indication in the car of the relative positions of the two sensors. That level of "feel" and understanding of the car is what sets professional drivers apart.

Gentleman Drivers

At many professional levels of motorsport, in particular within endurance racing, you will often find what is known as a "Gentleman Driver" within a team. Gentleman drivers are not professionals, but that doesn't necessarily reflect on their skill level. Gentleman drivers fund their own seat in the team and often bring additional funds and sponsorship.

Gentleman drivers are found in almost every professional championship of motorsport, from Formula One to the World Endurance Championship.

Gentleman drivers tend to be the most difficult to generalize. Sometimes they are extremely friendly and simply happy to be able to race. Other times, you will find a "My Money, My Rules" attitude which can seriously impede the team's ability to compete effectively.

If working with a gentleman driver, always bear in mind that they may effectively be signing your pay check. Be nice, play nice and they will most likely reciprocate.

Professional Relationships

The relationship between driver and crew is essential to championship success. The driver should have the confidence and faith in the crew that they have set the car up well, and that it is above all else, safe! A driver should trust the judgement of the engineers that are running the car, and there should be a mutual respect between everyone that contributes to the race. If your driver doesn't trust you, this will cause friction in the team.

Equally, engineers and mechanics, the people running the car, should trust the driver. They should have confidence that the driver will not take unnecessary risks, will race to their absolute best ability, and will extract all of the performance from the car. A driver that consistently brings a car back battle scarred, damaged and broken, will not be popular amongst the crew, particularly if those battle scars are not indicative of success.

The combination of driver and race engineer, and even driver and entire crew, is something that can be carried from team to team. As drivers progress in their careers, they will often opt to bring their race engineers and their mechanics with them to each successive step on the motorsport ladder. A good relationship with a talented driver can be a gateway to progression within the industry. Keep this mind when working in the lower rungs of motorsport – the people you meet might take you with them if they start moving on up. This is discussed in more detail in Networking.

Chapter Seven

OTHER THINGS TO KNOW

"Real knowledge is to the know the extent of one's ignorance."

– Confucius

There are countless elements to working in motorsport that I was not aware of starting out, and many of my friends were also unaware or misinformed. This chapter is by no means extensive, but should provide a few helpful pointers for your first trip to a race track as a professional. This chapter should work as a standalone checklist to be read over before and during your time in a trackside environment. It is not necessarily about career progression, but about surviving and remaining comfortable.

Food and Drink

Food at any track will vary depending on the type of event and the team or championship. If you are attending a test day, expect very limited catering facilities compared to a race event which will likely have dedicated catering companies for the teams.

I highly suggest that you eat whenever the opportunity is given. If you have a quiet moment, use it to fuel yourself. Quiet moments do not last, and you cannot guarantee when you will next get the opportunity to eat.

Also be aware of others in the team who do not have the same quiet moment you do. Offer to bring them food over (or just bring it and leave it on the side). It is always appreciated by busy mechanics and engineers.

The same goes for drinks. Tea and coffee may be a staple of many race teams around the word, but when working hard, be sure to keep hydrated at every opportunity. If making tea or coffee, offer it to your colleagues and bring it over. Bring cups of water to the mechanics if they are breaking a sweat. (Warning – do not put water anywhere that it can be knocked over. You will not be popular if you create more work in clearing up, or worse, damaging tools and equipment due to water ingress!)

The quality of food varies considerably at the track, and will range from a burger from a van to a three course set menu that could easily be served in top flight restaurants. Whatever is on offer, you cannot afford to be overly fussy. Eat what's there and carry on working.

Clothing

Pitlanes are notoriously cold during winter months. The openness of the track and garage blocks channels bitter winds at you all day, every day. The trick here is to wear layers – vests, t-shirts, jumpers, jackets and coats can be built into an adaptable outfit for changeable circumstances. If you're working on the car, chances it will be rather warm – you won't want all the layers then! But whilst the car is out and you are waiting, things get chilly! Consider having a hat and gloves in your bag during the winter months to cover those bitter January test days.

In contrast to this, there is very little shade in a pitlane during the summer. Again, layers to be adaptable, and shorts if they are allowed under team kit rules (I have found pants with removal leg bottoms to be particularly useful). If sunhats form a part of the kit, grab one and use it. This coupled with a good pair of sunglasses will help prevent aching cheeks and brows from squinting in the sun all day.

On the subject of sunglasses, make sure they are comfortable. You may end up wearing them for hours at a time and you will likely have a radio headset on as well. This can create pressure points above the ears which can get quite painful. You will find that Oakley sunglasses are preferred by a large number of people in the motorsport industry. Don't feel obliged to follow suit, but I do recommend them to people. They tend to be lightweight, comfortable and durable; three key requirements of anything you will be wearing for days on end.

Travel

Depending on the path your career takes, you may undertake extensive travel throughout a race season. If you are fortunate enough to get the opportunity to experience the countries and cultures in which you find yourself, I urge you to. You will find that a great deal of your time in each location is spent at the track, or at the hotel. It is quite rare to get a day to yourself in which to explore, so if the opportunity arises, take it.

Don't expect to have days of free time in each destination. Once the race finishes, you will most likely be flown back out of the country. Some teams will allow you to extend your visit to locations by postponing your return flight. Any hotel and food costs incurred will be at your own expense, but it can provide an excellent opportunity for seeing the world with your flight costs already covered. This is of course dependent on your commitments and timings for the next race or obligation, but a day or two is usually possible at some point of the season. Remember though that these extra days spent there will be taken out of any vacation allowance.

A truly international championship, such as Formula One or World Touring Cars, can mean weeks on end away from home, but also limited luggage capacity. Packing for such an event is different than a national series.

Be clear beforehand about what your luggage allowance is. If hand luggage only, you will need to master the art of packing light. If you can check hold luggage, consider if it is necessary and weigh up the additional hassle of checking and collecting the luggage against the time you are travelling.

A large amount of your allowance will be taken up with team kit, so allocate space for this first. Clean underwear is obviously essential, and I recommend bringing four pairs for every three days away as a minimum (assuming you do not have laundry facilities during that time). I always carry spare underwear, socks, and a spare team kit top in hand luggage, along with a phone charger. You can survive a remarkable amount of time with this setup in the event of hold luggage getting lost.

Particularly when starting out in the industry, travel will seem like an excellent perk to the job. Travelling the globe for work is a dream for many people, and it can be a fantastic experience. You will experience cultures and meet people who you would otherwise never have had the chance to. It is worth keeping in mind though, how extensive and long term travel can impede on your home life. It's a balancing act, and one that a lot of engineers grow out of as their career progresses. With families and mortgages, and the responsibilities of life adding up, many engineers will settle into national championships later in their careers, having already experienced the international stage.

Representing your team

It is important to remember that when you are at an event, you are representing your team. If you come across as professional, then the entire *team* will come across that way as well. If you are seen as unprofessional, this too will reflect on the team. Be aware of how your behavior, appearance and attitude might impact the team's reputation in the paddock. If you pull it into disrepute, you may not be asked back for the next race.

The same goes for the social events that frequently happen towards the end of the day. Wearing team kit in bars and clubs should be avoided, and whilst drinking alcohol is absolutely fine, be aware that again, you represent the team you are competing with. Your behavior should not give cause for concern to your team managers.

On the subject of alcohol, at most events there will be certain team members who are required to undergo a breathalyzer test before the competition each morning. If you are on this list (and it is *not* just the drivers) be aware of your consumption the night before.

Additional Equipment

Sun Cream – This is something I failed to bring to my first test day and regretted it for the following two weeks. The sun can be relentless in a pitlane as mentioned earlier and adequate sun protection will save you days of discomfort. You are not at a race track to get a sun tan – arms and face as a minimum should be sun creamed up if you suspect a hot and bright summer's day.

Notebook and Pens – I always carry a small moleskin notepad and a reliable mechanical pencil. The moleskin notepad is hard wearing and the mechanical pencil does not smudge in the event of rain. These fit nicely in a midleg pocket of a pair of Dickie's pants and are easily accessible. Making notes will become critical to your tasks within the team and allow you something to reflect upon during post-event briefings. Everyone I work with, mechanics, engineers, team managers, officials and drivers, all of the *successful* ones keep notes. You won't remember everything later; I promise you that!

Permanent markers – Carrying at least two permanent markers of different colors has proved itself valuable on numerous occasions. I have used them for marking the position of sensors, nuts, bolts, wiring, torque settings, tires and many other items on the cars I have worked on and around. I have friends who use paint pens or torque seal for similar purposes. The advantages of these are they can be more resilient to the motorsport environment and you have a huge array of colors to choose from. This means you know if *you* have marked something and not someone else with the same blue marker pen!

It is also useful to carry a fine tip marker just in case your driver is without one when he or she is asked for an autograph.

A Watch – Any motorsport event should run like clockwork. Things will happen to the second, especially when they are being broadcast on television. An accurate and reliable watch will help you keep on schedule. I'd suggest leaving your expensive Swiss watch at home Something robust, such as a Casio G-Shock will withstand almost anything you can through at it. When working to TV schedules, having "radio watch" that updates to the exact local time will be extremely beneficial.

Lanyards - You don't want to lose your track pass or ID, so a strong, good quality lanyard will help here. A lot of the time passes will be issued with one, but it never hurts to carry a spare.

Dealing with the Media

Most motorsport events these days will be covered by the media in some guise. Be aware that reporters do not always make their presence known, wear press badges or carry voice recorders. Any fan that approaches you in the paddock should be treated with suspicion, and you need to be sure of what you can and can't make public.

If you are in any doubt, say nothing, claim confidentiality and refer them to the PR team. Reporters can be sneaky and devious, and nowadays, anyone with a twitter account can ruin your day quickly with a little insider knowledge.

Unfortunately, like most things, you cannot split this section into black and white. As a team member, you are still representing the team, regardless of whether the person approaching you is a fan or reporter. Treat them with respect, politely decline to comment, and carry on with your day.

Most events will have "official" media personnel who know who they can and can't talk to in each team. They will seek out drivers, principle engineers and the PR team for their information. Anyone coming straight to you is a chancer. Be wary. If it is just an over enthusiastic fan, remember that the fans bring in the money – they indirectly pay your wages so be nice!

TV crews can become quite intrusive, and unfortunately, if you are having a bad day, they are all the more likely to be close by. The professional camera men and women know the limits of their intrusion to a garage and if they are slowing you down or becoming an obstruction, you can ask them to move. So long as this is done professionally and politely, there shouldn't be a problem. Being able to filter out a camera that is 12 inches from your face whilst trying to resolve whatever issue it is that has made them interested can be difficult, but it's something that will develop with time. Don't acknowledge the camera is there unless you are being interviewed. Continue doing your job, maintain your professionalism, and bask in your 30 seconds of fame. If you do this, they will soon get bored and move on to another shot.

On the same note as talking to reporters, do not mention anything on social media that could be construed as giving away secrets. Unless the reason for a DNF has been released, you have no right to say it to your friends on Facebook. Photos of the car, the team, the trucks or anything else that makes up the team are not your property. Do not publish photos without getting permission from management or PR teams. The cars in particular have a huge amount of politics behind them with sponsors and manufacturers. On top of this, photos from the wrong angle or of the wrong component can give away team secrets and competitive advantages.

Parc Fermé

Parc Fermé is sacrosanct. Following a race, cars are put under "parc fermé conditions" which means no one can touch the cars other than the scrutineers.

If you so much as open the bonnet or door (after the driver has got out) you are risking disqualification from the standings. You don't want to be the person to make that mistake; leave the car alone until you have been told it has been released.

Raw Carrots

I'll finish this chapter on a rather obscure point. If you find yourself having to ride as a passenger in a car for any reason (to resolve traction control issues for instance), then I recommend eating raw carrots beforehand. This trick was mentioned to me by a lecturer at university and it really does work. It helps subside motion sickness and also the nausea that can be induced by the vibrations of the engine and chassis. Give it a go if this is something you suffer with.

Chapter Eight

NETWORKING

"Your network is your net worth."
- Unknown

Who you know, who knows you, what you know and what people know you know is what defines your path through the motorsport industry.

You might need to read that statement a few times for it to make sense – it was tricky enough just to write it down! This statement is the single biggest driver behind career progression within the motorsport industry. Let's take the sections one at a time...

Who You Know

Who you know is by and large the most important factor to achieve success and progression within the motorsport industry. Once you have secured your first job, which will most likely be through the usual application process mentioned previously, you must start building a network. Networking in motorsport means you are aware of new opportunities before the masses and can pursue those opportunities without the competition that comes with a public recruitment processes.

Throughout your career you will cross paths with many people. Some of these people will be household names; drivers, engineers, team managers or officials who you will have heard of before. Others will be enigmas, people who you have no idea who they are or what they do. The suggestion here is to make a point of meeting everybody you can. You do not know who everyone is; they may be the power behind the throne of a championship or team. They might be a nobody now, but become the next big name in the industry in the years ahead. If you make a point of meeting people, not being shy (and equally not being overbearing), then your network will grow in size and, more importantly, value.

Make a point of going along to industry events, trade shows, championship and team social gatherings; basically anything that will allow you to meet the current and future names of the motorsport industry. Introduce yourself to team sponsors if the opportunity arises. Thank them for their product or support and ask more about their business. Dig in to their interest in motorsport and start a rapport. - you never know where they will put their sponsorship budget next season!

Who Knows You

Following closely on from who you know is who knows *you*! Shaking hands with a driver after a race or a thirty second "congratulations" to a winning engineer does not mean they *know* you.

When meeting new faces in the industry, it is important to make a good impression. Making yourself memorable without coming across as overly confident or arrogant is a fine art, but one that you should endeavor to master. Being courteous, funny, polite and intellectual means you can truly count that person within your network.

Your network is also not self-sufficient. You cannot rely on contacts creating contacts. Just because you know Mr. Smith, and Mr. Smith knows Mrs. Johnson, does not mean *you* know Mrs. Johnson. If you want to gain contacts through contacts, which most people do, then you can always ask for an introduction. Express an interest in an accolade, achievement, career choice or publication and use that to strike up a meeting at the next opportunity.

Also, once you have someone in your network, nurture the relationship. That doesn't mean inundate them with phone calls, emails, texts or letters, but if the opportunity to strike up a conversation arises, do so. You might find them at the same race or event, or at a trade show or industry conference. Make the effort to remind them of who you are. Keep these relationships fresh and you will soon be the owner of a large and healthy network of valuable contacts.

You can also get your name out in to the industry by submitting articles of interest to trade publications or even keeping a blog. If it is publicized enough and of good quality, your name will become known to a lot more people than you will ever have the time to meet in person. It also becomes a great talking point when meeting people for the first time.

Just a final note on getting to know people. I do not intend this chapter to encourage fake and one-way relationships. It is important to demonstrate your value to them as much as acknowledge their value to you. If you are clearly only after them as a way of progressing your own career, you will soon find your network shrinking faster than it grows. Be sincere, and these relationships will be mutually beneficial.

What You Know

What you know is by nature an entirely open ended subject. When you start out in motorsport, you may feel you know everything there is to know about your area of expertise. I do truly hope that by this stage of this book that mindset has been quashed! Fresh out of university, college or an apprenticeship will no doubt give you a solid foundation on which to build a wealth of knowledge, but you are far from an expert in anything.

A willingness to learn is a key attribute that employers and teams will look for, so it is important to acknowledge that you still have so much to learn.

The body of knowledge that you have inside your head is what will set you apart from peers and competitors, and will be absolutely key to career progression within motorsport.

Whenever you find yourself in an unfamiliar situation or tackling a new problem, do whatever you can to absorb new information and knowledge. Spend time with more experienced colleagues. Being on the frontline in problem solving is the perfect way to expand your knowledge and prove yourself as a capable engineer. And the more you know, the more valuable you are. Your team will want to keep you, and the competition will want to steal you. Working to increase your value in this way will pay dividends throughout your career.

What's more, you will become more practiced and learning and problem solving, which in themselves are excellent and priceless skills to any engineer.

What They Know You Know

Proving and demonstrating the knowledge you have is a tricky business. Nobody likes a "know-it-all" but equally, an obscure snippet of knowledge could mean the difference between a win and a second place. Depending on the role you have within the team, you will discover different ways of making your knowledge known.

Passive knowledge demonstration is the most common form, and is extremely effective. Unfortunately, this means there is no "in your face" forcefulness of showing how intelligent you are. You demonstrate your abilities, and your intelligence, through the work you do. These guys will turn up, do their job and do it well. Their results speak for themselves and they tend to be the "cool as a cucumber" under pressure people.

The other possibility is those people who are vocal and bullish with their opinions and knowledge. In some industries, this would be seen as counter-productive and unprofessional. In motorsport, less so. Decisions often need to be made very quickly and communicated instantly to mechanics, drivers or other team members. There is no problem in being vocal with knowledge so long as you are correct in what you say.

The best advice for starting out would be let your results speak for themselves to start with. Wait until you have a greater knowledge base than is taught in a lecture theatre before speaking authoritatively on an issue or subject. Perhaps even seek out the vocal engineers and express your opinion and knowledge to them. See if they agree enough to make it known amongst the entire team. You will often find you will be credited with the knowledge. (Warning – you will almost certainly be credited if it turns out to be wrong!)

The best engineer in the world will struggle to progress if he does not have a proven track record (no pun intended). Starting out in this industry, you are fortunate – nothing is expected of you. You have no track record. Getting a job is a blessing as it allows you start developing one.

If you thought getting your first job was difficult, your second job could prove to be even harder. You will have to show a history of success, of value added, in order to secure your next job.

Your Reputation

You will probably have heard the expression "Your reputation precedes you". It certainly does in motorsport. My reiteration of this point throughout the chapter, and indeed the book, is intentional. There are several aspects to your reputation, and "What you know" is just one of them

With a strong and growing network, and a good history of success you will be well on your way to securing a good reputation. But there is more to this industry than who you and what you know.

If someone in your network is asked "what do you think of him/her?" you want the answer to be entirely and instantly positive. You want to be someone who is a pleasure to work with and who is a real team player. Someone who maintains their cool under pressure, has a professional demeanor, but also someone who is sociable and popular within a team.

A reputation like this cannot be learnt from any book (including this one!). It is something you must consciously work on until it becomes natural.

In any walk of life, a bad reputation will spread far more quickly than a good one. A bad reputation self-replicates as opinions are offered by others without being prompted. Think how many times have you moaned about a colleague or classmate to someone without them ever asking? It is human nature to vent frustration and annoyance. But rarely will you offer positive observations about someone without first being asked what your opinion is. This fact alone makes building and maintaining a good reputation far more difficult then losing it

Having a good reputation will put your name to the top of the list and at the front of people's mind when they are asked for recommendations. This is how you progress in motorsport, and becomes even more critical if you begin working for yourself (see the next chapter). Prove yourself to get recommended.

Chapter Nine

Going Solo

"Success is a staircase, not a doorway."
— Dottie Walters

Starting your own business and being your own boss is the Holy Grail for a lot of people. It really is exciting to choose where and when to work, who to work with, and ultimately dictate your salary. Excellent! You might think…

I cannot encourage you enough to have this as your goal. Once you are your own boss, you will experience the real freedom and broad nature offered by the motorsport industry. This does, like everything else in this book, come with a few caveats.

Cash Flow

Chiefly, if you cannot find work, you do not get paid. This can cause a huge amount of stress and anxiety, particularly if you have mortgage or rent payments, a family to support, or a retirement to save for. The work for a contract engineer in motorsport can be described as tidal. During the race season, you will most likely be immensely busy. Race weekends will eat up 5 days each week (Thursday through to Monday, inclusive), and you may even be involved with multiple teams and championships. You will be issuing invoices on a weekly basis and all will be well. Life is good.

However, come the end of the season, your services are no longer required. Testing will be several months away and there will be no races for maybe half the year. Suddenly, your income drops to zero.

This however, is not necessarily terminal. All it takes is some proper forward planning and you can ride this inter-season storm pretty well.

One thing several contractors I know do, is to run a secondary business on the side. Perhaps one that will be busy during the off-season months. Parts sales, design contracting, journalism and driver tuition are some potential money makers to be explored. Plan to be out of work for 6 months of every 12 and you should survive without too much of an issue.

I also recommend that during the off-season time you don't allow relationships with teams to go stale. Keep in contact every few months so that when they are looking to swell their ranks for the next season, your name is high on the list.

If you can secure a contract for the next season early, fantastic! You're one of the lucky ones. Don't be surprised if getting a contract with the same team year on year comes down to the wire each time. Also, beware that it has been known for key sponsors to pull funding at the last minute leaving the team (and your contract) high and dry; have a back-up plan wherever possible.

Image

As your own boss, it is important that you give off a professional demeanor at all times. As mentioned previously, reputations within motorsport will make or break an engineer. Be courteous, be polite, be fun, but be professional! Your professionalism will help to guarantee repeat custom and also help you keep a high number of back-up customers.

A professional image can be reinforced with simple and cheap things such as a smart website, sensible business cards and even branded clothing for meetings and trade shows. These things show an investment in yourself and give potential clients a means of judging you as a professional before ever having worked with you.

A frequent mistake that is made by new solo engineers in motorsport and a mistake I feel I made looking back, is going at it alone too soon. As a contractor, your services will demand a premium and your skillset will be expected to reflect that cost. When starting out it is easy to feel that you could do this on your own. However, before you start, ask yourself two questions:

If I didn't have the backup of the team or company I currently work for, can I still hold my own?

As a team member, your mistakes are much more tolerated. If you screw up you might get shouted at, but that's about it. As a contractor, if you screw up you will be told to leave and not come back. Contract terminated. Income gone. As a contractor you cannot afford to make the same mistakes that you can do as an employee.

Does my reputation allow me to demand a premium?

If your reputation is non-existent (which is different to it being "bad") then when you present a quote for services, you will get questioned. I have worked with some excellent engineers in the past, some of whom are the very best at what they do. When they get asked to work with a team, there is no discussion on day rates. They demand a premium, the team know that and due to reputation alone, the team are willing to pay it. When you start out solo you won't be at that level, but you will want at least a basic level of mutual respect between you and your customers. Starting from zero is a fool's game.

Keeping on Top of Things

As your own boss, your time management becomes all the more important. It is far too easy for time to be spent on clients, customers and speculators who in actual fact bring in little or no value. The best system for ensuring you do not waste time on fruitless endeavors is to redefine how you are contacted. I have three means of contact: a website, an email address and a phone number.

The website is accessible to everyone. Anyone can visit and send a message requesting my employment or assistance with a project, race or championship. Obviously, this can result in a large amount of inbound-email, so I have this filtered by email rules and only respond to these requests a couple of times a week. This reduces the amount of time whilst at the same time removing the urgency induced stress that can come with high message flow.

My personal email address is sacred and shared only with clients and customers with whom I have an existing relationship, or a fruitful one in the pipeline. My business cards are not handed out to everyone and if people really push me, I can direct them to the website or an "info@..." email address instead. This means emails that come through to me are meant for me and time responding is time well spent. I do not however, use email as an instant messaging service. I will respond to email a couple of times a day, but an immediate response is very unlikely. You will find that anyone emailing you rarely expects an instant response and that it is only a pressure generated by ourselves that compels us to check and respond as frequently as we do.

Finally, my personal phone number. This is given out to clients I am actively working with, at a race track, or clients with whom I have had a successful relationship with in the past. If people phone me, they can expect an answer. If I happen to be in a meeting, I will reply at the earliest opportunity. Phone calls tend to be one of two things; either an emergency that requires a fast reaction, or a promising and exciting opportunity that has just come up. No one calls for a chat.

Marketing

Now that your time is free from responding to emails for most of the day, your focus should be on securing your next job. Marketing yourself is a difficult thing to master for most people. I struggle with it even now. You must assume that work will not come and find you, and that you must go and find it. Spend some time calling friends, colleagues and clients and asking if there are any opportunities. Also, be sure to increase your network (as discussed previously) by visiting trade shows and tracking down new teams and companies who might require your services.

Maintaining a professional website as well as some social media presence will certainly help in securing the professional look you want. You should have personal and company LinkedIn accounts as a minimum and a presence on twitter can be a quick and easy way to share previous experience (remember to check with teams if they are ok with you publicly displaying information!)

When visiting trade shows, spend some money on clothing with the company logo on. This again helps secure a professional image and will stop you appearing as a speculator when you approach teams and companies. Couple this with a professional business card and you will really be helping yourself!

When negotiating contracts with teams, see if you can have your company logo as a sponsor on the car. Make it clear that it doesn't have to be large and offer a discount on your rates if necessary. Being able to show that affiliation helps when approaching future clients.

I do not want to put anyone off working for themselves in motorsport. You will gain a fantastic insight into the spectrum of the motorsport industry and can potentially see some excellent returns for your time. I also don't want anyone being blasé about it and getting burned. Motorsport is a dog-eat-dog world. It is brutal and survival of the fittest is true at all levels. With the right preparation, and the right attitude, you will make it!

Chapter Ten

So, in Conclusion

"The only way to do great work is to love what you do."
– Steve Jobs

I hope that by reading this book you have gained an insight into what motorsport is really like. It is a diverse, exciting and complex industry, one that is full of passion and one that is notoriously difficult to succeed in.

I truly love working within motorsport, and everyone I work with, past and present, says the same. It is a tough industry and one that takes few prisoners, so those that are there day in, day out, they are there because they love their job.

There will have been aspects of this book that were perhaps off-putting to new engineers coming into the sport. There may have been things you disagree with or things that you will *come* to disagree with as your own career progresses. All I can do is give you a heads-up to the things that aren't all that well known. Those little caveats that take some people years to realize and others never actually do.

When I started out in motorsport, I was shocked by how little I really knew about an industry I had strived to break into for years. The image in my head of how motorsport worked, of how it functioned, was far removed from the reality of racing.

That is not to say that I had my dreams crushed. Not at all! I had simply never considered the things that were not taught throughout my education and never shown on the television. My eyes were opened very quickly as small mistakes I made at the start of my career were repeated by the new recruits that joined after me.

I have come across people who worked for years to get into motorsport only to find that it is not how they envisaged it. The superyachts and private jets were replaced by 18 hour journeys by road and economy tickets all over the world.

My biggest fear with writing all of this down is that it will put off some of the best new talent that is currently coming through the education system. This is far from my intention. What I want to achieve is for those new engineers and all the others who may have found this book useful, to be better prepared and even more determined to really make it in motorsport.

The fact of the matter is that whatever industry you work in, you will have ups and downs, good days and bad. In motorsport, the chasm between these two can be extreme. On your down days, you might see your car go up in flames or good friends get hurt. That is an unfortunate reality of the industry. However, on your good days you will experience comradery, a real sense of achievement and get a taste of victory. You will be part of a winning team and the drive to succeed will be reignited in you.

Appendix

Recommended Reading

The books below are a selection of books I personally recommend, or have had recommended to me. I am not paid to endorse anything below, nor do I claim any of the below as essential. You won't need to read every page of all of these books, however, a selection of them will help you through a lot of what is thrown at you whilst working in motorsport.

Engineer to Win by Carroll Smith – Considered a must have reference book for anyone trying to succeed in motorsport. The writing is concise and you will learn a huge amount of the required knowledge to get by.

Speed Secrets Series by Ross Bentley – An excellent way to understand the dynamics of a racing car and decipher the language being used by your driver. Speed Secrets is relevant whether or not you are the driver.

Race Car Vehicle Dynamics by Milliken and Milliken – The bible when it comes to vehicle dynamics, covering every aspect of the dynamics and kinematics of the chassis. This also delves into the spooky world of understanding tire behavior. A perfect reference book, especially whilst studying.

Race Car Aerodynamics: Designing for Speed by Joseph Katz – Understanding aerodynamics is not something that happens overnight. Katz aids students and professionals alike into getting a grasp of how to understand and improve the aerodynamics of your race car.

Engineer in Your Pocket by Carroll Smith – A short but genius flipbook that really helped me out at the start. On one side is Cause followed by Effect, and on the other, Effect followed by Cause. Want to know how a change will affect the balance of the car, or how to counter an effect being described by the driver? This pocket manual is the perfect trackside companion.

Making Sense of Squiggly Lines by Christopher Brown – A fantastic ground-up introduction to data analysis and system engineering. Perfect as a reference book for any engineer reliant on data to improve their car.

Hillier's Fundamentals of Motor Vehicle Technology by Alma Hillier – Gaining a fundamental understanding of the systems on a car is critical to being able to improve them. Hillier helps the reader start from nothing and work up to an extensive body of knowledge on which to build.

Introduction to Internal Combustion Engines by Richard Stone – Understanding engines is a lifetime's work. Fortunately, Stone has condensed years of experience into an easy to digest, comprehensive and detailed text book. Not ideal for carrying to the race track, but excellent for studying.

Hands-On Race Car Engineer by John Glimmerveen- This book touches on every aspect of a racing car, from design principles, to setup changes, to driving. Everybody can learn a thing or two from this book and it will be valuable from the very start of your career.

The Obstacle is the Way by Ryan Holiday – This book is not written with motorsport in mind, but provides fantastic guide to maintaining the right attitude in the face of hardship. You'll have bad days in motorsport, and you'll have tough decisions to make. Holiday helps out!

The Four Hour Work Week by Tim Ferriss – Running your own business can quickly consume your entire life leaving no free time whatsoever. Ferriss details some simple tips and tricks that help you get the most out of your business and keep it growing.

www.startingonpole.com
© Torotex Engineering

Printed in Great Britain
by Amazon